嘻哈口语

你说东，我说西

Everyday Dialogues on Cultural Differences

[英] Nick Stirk 著

张满胜 译

外文出版社
FOREIGN LANGUAGES PRESS

Introduction
序 言

These dialogues will make you laugh, make you cry, make you think, but most of all they will make your English better. They are special in three ways. First, they have been written by a native speaker who has had several years experience in China teaching at top universities. They are full of up-to-date natural expressions used by native speakers, which means that you will be using the same language that they speak. Second, there is an emphasis on coll-ocations, phrases and expressions. This is because two, three, four and even five-word collocations, phrases and expressions

make up a huge percentage of all naturally-occurring text, spoken or written. Estimates vary, but it is possible that up to 70% of everything we say, hear, read or write is to be found in some form of fixed expression. Therefore, it is vital to realise that this is the way we store vocabulary in our lexical minds. We are more likely to memorise and use phrases rather than single words. So get rid of word lists and replace them with lists of phrases. Third, and most important of all, culture involves everything we say, do and think, so it's very important, not only in our own culture but also when we interact with another culture, that we develop a culture consciousness. Knowing the differences and allowing for them is very important when we meet someone from another culture, whether in their country or our own.

I wish to acknowledge the encouragement and assistance provided by my publisher Cai Qing. I would like to thank my English majors at Beijing University of Posts and Telecommunications for their help in road-testing these dialogues and for making many valuable suggestions. I would specially like to thank Li Yang for her suggestions on various dialogues. I am truly grateful for her insight and knowledge of intercultural communication. Finally I would like to thank Tori Waters-Wang and her husband Wang Qing for contributing some of the dialogues and to Zhang Mansheng for his skilled translation and for his ability to transfer my sense of English humour into Chinese.

本书中的对话会让你哭，让你笑，让你思考，最主要的是会让你的英语有所长进。特别之处有三：其一，本书的作者来自英国，并且拥有多年在中国一流大学的教学经验。对话中充满了最时尚的表达方式，也就是说，你将会学到现在英国人最流行的口语。其二，对话中突出强调了词组、短语和表达法。因为在任何课文中，无论是口语还是书面语，由二、三、四、甚至是五个单词所组成的词组、短语和表达法都占据了主要部分。在我们日常的听说读写中，固定表达法的比例可能会高达70%，这就是我们大脑储存词汇的方式，意识到这点很重要。与单个的单词相比，我们更倾向于记忆和使用词组。所以，在学习英语时，不要只记单个单词，而应该记忆词组和短语。第三，也是最重要的，我们日常所说、所做与所想都会涉及到文化，因而在我们进行跨文化交流过程中，要有文化意识，这点是非常重要的。不论我们身在异国还是在自己的国家，当我们遇到外国朋友时，了解并尊重彼此之间的文化差异，是我们进行良好交流的基础。

非常感谢我的出版人蔡箐小姐给予我的鼓励与帮助。感谢北京邮电大学英语系的同学为本书中的对话进行了实际演练，并提出了很多宝贵的意见。我要特别感谢李杨对对话提出了许多中肯的建议，以及她对跨文化交流的丰富见解。最后，我还要感谢 Tori Waters-Wang 以及她的丈夫 Wang Qing 先生，为本书贡献了许多精彩的对话。同时，还要感谢张满胜先生的出色的译文，准确传达了该书的幽默风格。

How to Get the Most out of These Dialogues 使用指南

Role play dialogues are excellent opportunities for learners of English to practice English because they give students the chance to assume a role that is not their own. They can be anyone they want. It does not matter if you make a mistake because it is not you who makes it. It is your character who does! You can totally assume the character and personality of your part and play it to the hilt and beyond! Your voice and actions can be completely over the top and yet it does not matter because it is only acting. Role playing is a safe and enjoyable way to learn English. It is a good idea to find a partner to practice the following dialogues with.

Tips on role plays

1. Don't be afraid to act the part
2. Try to get into the character
3. Explore different voices
4. Use appropriate movements and gestures
5. Change parts so that you can play a different character
6. Don't be afraid to play a character of the opposite sex

Studying spoken English really can be easy, even in a fast-paced and demanding world. Take this book with you on the go. As each dialogue is independent of the others, you can study them piece by piece. Take the 15 minutes you're stuck in traffic to read a topic. Carry it with you on the subway. Squeeze in a few minutes here and there. If you practice one dialogue a day then at the end of 50 days you'll be amazed at how your English has grown. Your confidence in speaking will be sky high and when a foreigner appears you will have plenty to talk about!

对于英语学习者来说，在英语会话中进行角色扮演是锻炼英语水平的一个绝好的机会，这让学生们有机会去担任一个完全不同于自己本人的任何角色。即使犯了错误也无所谓，因为并不是你犯的，是你扮演的角色犯下的！你完全可以设想自己所扮演角色的性格特征并将其发挥到极至。你的声音、动作都可以很夸张，这并没有什么，不过是在演戏罢了。角色扮演是一个安全并且令人愉快的英语学习方式。找个搭档与你一起练习书中的对话，是个不错的主意。

角色扮演小贴士

1. 表演时不必有任何顾虑
2. 努力融入角色
3. 语言表达要抑扬顿挫
4. 使用恰当的肢体动作和手势
5. 进行角色互换，以便尝试不同角色
6. 不要羞于扮演异性角色

学习口语其实很简单，即使在这个快节奏、竞争激烈的社会里。随身带着这本书，书中每段对话都是相互独立的，你可以逐一学习。路上堵车15分钟，你不妨就来读一个话题。坐地铁时也带着它，四处挤点时间就够了。要是你每天练习一篇对话，50天后你就会惊讶于自己英语的进步。你在讲英语时会非常自信，再遇到老外的时候，你就不会无话可说了。

目录
Contents

1 约会

Dating

▌▌ **Background Information** ▌▌

Humans are emotional creatures created with a vast capacity to love and to be loved. This need to give and receive love is fundamental to life. We are born from love in order to love. With love the impossible is nothing; yet without love nothing is possible. Love makes the world go round, makes your head spin, makes your heart beat faster and makes your day. There are no limits to love. A grandmother can love a baby and a baby can love a grandmother. An Englishman can love a Chinese girl and a Chinese boy can love a Korean girl. Love can be timeless, eternal and never ending. It can also be brief, spontaneous and passing. Love is cool. Love is hot. But love can never be boring. The eyes sparkle, the heart flutters and the toes twinkle.

　　人是有感情的动物，生来就有爱与被爱的巨大的能力。给予爱与接受爱成为生命的必需。我们因爱而生，为爱付出。有了爱，一切皆有可能；没有爱，一切皆不可能。爱让世界转动，爱让你头晕目眩，爱让你心跳加速，爱让你幸福快乐每一天！爱无所不能。祖母可以爱孙儿，孙儿也可以爱祖母；英国绅士可以爱上中国姑娘，中国小伙儿可以爱上韩国女孩。爱可以是长久的、永恒的，也可能是短暂的、稍纵即逝的。爱可以是一丝清凉，也可能是一团热火，但爱不会令人乏味。爱让你明眸闪亮，内心悸动，双脚雀跃。

背景信息

Dialogue 1

Helen: What's your opinion about **cross-cultural love**①? Would you ever date a local guy?

Tori: To be honest, I wouldn't. I don't find Chinese men attractive, but not because we are different cultures. I mean, I'd date someone from another Western country, no problem. How about you, have you ever dated a foreigner?

H: All the time! I love foreign men! Especially the older ones because they know how to treat a girl right. **Wine her**②, **dine her**③, buy her romantic gifts. I think foreign men are simply more romantic than Chinese men.

T: I wouldn't say that's strictly true, especially nowadays as China is becoming more open and Westernised. I've met lots of romantic Chinese men.

H: I know that Chinese men look after you until the wedding but then once they are married they stop and focus on their career.

T: Yes, I think that's true. But as I said, I wouldn't do it because I don't really go for Chinese guys. And I find they don't go for me either. Maybe my size puts them off!

H: Yes, I've noticed that Western girls are a lot bigger than Chinese girls.

T: And most Chinese men are smaller than Westerners so they don't want to date someone who is taller than they are.

H: Maybe I should introduce you to Yao Ming!

习惯用语 1

① cross-cultural love: 异国之恋，跨国婚姻（来自两个不同国家的人彼此相恋）
② wine sb.: 约某人出去一起喝酒
③ dine sb.: 约某人出去一起吃饭

H: 你对跨国恋怎么看？你会跟这里当地的小伙子约会吗？

T: 老实说，不会的。我觉得中国男人不怎么吸引人，并不是因为文化差异。我的意思是，我会和其他西方国家的人约会，这没什么。你呢？你跟老外约会过吗？

H: 我一直都和老外约会！我喜欢外国男人！尤其是年龄大的，他们知道怎么疼女孩子。请她喝酒，约她吃饭，给她买浪漫的礼物。我觉得外国男人基本上都比中国男人浪漫。

T: 也不绝对，尤其是现在中国越来越开放，越来越西化。我就遇到过好多浪漫的中国男人。

H: 我知道中国男人在婚前总是很照顾你，一旦结了婚，他们就把精力都扑在事业上了。

T: 是的，我觉得这没什么不对。但就像我说的，我不会去和中国男人约会，因为我对他们不感兴趣。我发现他们对我也是如此。估计是我的体型吓到他们了！

H: 是的，我注意到了，西方女孩比中国女孩高大。

T: 而且大多数中国男人都比西方人矮小，因此他们也不想跟比自己高大的人约会。

H: 也许我该把你介绍给姚明！

Questions 1

1. Have you ever dated someone from another country?
2. Who do you think is the most romantic, Chinese or Westerner? Why?

Dialogue 2

Helen: I had a date with an older American man last weekend. He took me to a beautiful restaurant, we drank expensive imported wine, and then I went back to his massive, international apartment, where we had a romantic evening. Nothing like my local ex-boyfriend, he would take me to a Chengdu snacks restaurant and buy me a Beijing beer.

Abbey: Exactly! You said it, "expensive". Maybe your last boy-

Love 爱情

friend wasn't lacking in the romantic department, he was lacking in cash. He bought you all he could afford. Your situation is not about romance, it's about money. Anyway, how does your family feel about you having all of these foreign boyfriends? Does your mum think they are **good marriage material**①?

H: Actually, my mum is very happy with my dating foreign men. Since China has been open she's travelled to Australia, where she always dreamed of going, many times. She hopes that I find a rich Australian man who will help her to emigrate! What about you?

A: No way! A traditional Chinese mum would like her daughter to settle down with a traditional Chinese boy from a good family, with a good education and a good job.

H: Yeah, and if she has anything to do with it, she'll choose him! I'd rather choose my own lover because I love him rather than it being my parents' choice.

A: That reminds me of a joke. A boy kept bringing his girlfriends back home but his mother never approved of them. Finally, he took home a girlfriend who was just like his mother in every way.

H: So his parents approved of her and they got married?

A: No! His mother liked her but his father didn't!

習慣用語2

① good marriage material: 适合结婚的人

H: 上周末，我和一个年长的美国男人约会了。他带我去了一个漂亮的餐厅，我们对饮昂贵的进口葡萄酒，饭后，去了他宽敞的国际公寓，一起度过了浪漫的一夜。他可不像我以前的中国男友那样，带我去成都小吃店，喝北京啤酒。

A: 没错！你说了，"昂贵"。也许你的前男友并不缺少浪漫，而是缺钱。他给你买了所有他负担得起的东西。你现在关注的不是浪漫，而是金钱。你家人对你的这些外国男友怎么看？你妈妈认为他们是理想的结婚对象吗？

H: 其实，我妈妈很乐意我与外国人约会。中国开放后，她有机会去了几次她梦想中的澳大利亚。她希望我能找到一个有钱的澳大利亚人，帮她移民。你怎么看？

A: 不会吧，传统的中国母亲都想让自己的女儿与传统的中国男子结婚，这个男孩要有好的家庭背景，受过良好的教育，拥有体面的工作。

H: 没错，如果她真能自己选择的话，她也会选择这么一个男孩。我要选择自己的恋爱对象，我的爱远比父母的选择重要。

A: 这倒使我想起一个笑话。有个男孩总是带不同的女朋友回家，可是他的妈妈却总也不满意。最后，他带回家了一个在各个方面都与他妈妈极其相似的女孩。

H: 他的父母很满意，然后他们就结婚了？

A: 不是，男孩的妈妈很满意，可是他的爸爸又不满意了！

Questions 2

1. Do you like a boyfriend who buys you expensive things?
2. Would you marry someone your parents did not approve of?

Dialogue 3

Frank: You don't see too many East-West couples where the man is Chinese and the woman is a foreigner.

Nick: But when you do they tend to be married. I know of at least two Western girls who are married to Chinese and they seem very happy.

F: Yes, but you do see plenty of East-West couples where the man is a foreigner and the woman is Chinese.

N: And usually the woman is in her thirties. She is either divorced with a child or is a left girl.

F: A left girl? What's that?

N: They're called that because they're left behind and **left on the shelf**[1]! They're also known by their three H's—high diploma, high salary and high age! They're also known as the three S's—single, stuck and born in the seventies!

F: So because they find it difficult to get a Chinese husband they turn to Westerners!

N: That's right! Most Chinese men want to marry someone who is younger than they are and also less educated or at least they don't have a higher degree than they do.

F: You mention divorced women with a child. Why do they **go for**[2] Western men?

N: Partly because it's difficult for them to remarry a Chinese. As I said before, most Chinese men want to marry a younger girl and they don't want a woman with a child because education is expensive in China.

F: Well, so long as they marry for the right reasons then everyone is happy.

习惯用语 3

① leave on the shelf: 束之高阁，无结婚希望的（大龄单身女子）
② go for: 选择，对…有好感

F: 在东西方跨国婚姻中，你很少见到男方是中国人，女方是外国人的。

N: 也有啊。我就知道至少有两位外国女孩嫁给了中国人，他们看起来很幸福。

F: 这倒也是。不过，在东西方跨国婚姻中，更多的还是由西方男人和中国女人组成的夫妻。

N: 通常，女方已经 30 多岁了，她要么是离婚，还带个孩子，要么就是"无人问津"了。

F: 无人问津？什么意思？

N: 是因为她们都是大龄女青年，已经没有人愿意和她们交往了。她们以"三高"著称——高学历、高薪水、高年龄。她们往往是单身，难以找到对象，都是生于 70 年代。

F: 所以，她们觉得很难找到满意的中国丈夫，于是就把注意力转向老外！

N: 没错！大多数中国男士都想娶一个比自己年轻、比自己学历低或者至少与自己学历相当的女子。

F: 你刚才提到了离异且有孩子的女人。她们为什么要找老外呢？

N: 部分是因为与中国男人再婚有点儿困难。就像我刚才说的，大多数中国男士想找一个比自己年轻的女子，他们不想娶一个带着孩子的妈妈，因为在中国，教育的开销是很大的。

F: 嗯，其实，只要大家结婚理由正当，就会很幸福。

Questions 3

1. Do you know of any western girls dating Chinese men?
2. Do you know any left girls?

2 恋爱

Relationships

|| **Background Information** ||

A 2006 survey by Tom Online shows 68 percent of respondents would like to try a cross-cultural relationship when conditions allow. However, the survey also showed that 68 percent viewed cultural differences as the main obstacle in a relationship. As with any relationship the key to success is to have more things in common than are different.

2006 年，"Tom 在线"的一项调查显示，68% 的受访者在条件允许的情况下愿意尝试跨国婚姻。然而，调查也显示，68% 的人认为文化差异是两人感情中的主要障碍。因此恋爱成功的关键在于双方的共同点要多于差异。

背景信息

Dialogue 1

Teresa: What are some of the problems you've had since dating James?

Cassie: Well, we get stared at a lot in the streets which I really hate. James doesn't mind because he's used to the stares.

T: Why don't you like it?

C: Perhaps some people think I'm a bad girl who's **only interested in one thing**① but James isn't rich at all. He's an English teacher so his income is not much.

T: What did your parents think?

C: When I first told my mother she couldn't sleep for two weeks. When they finally met all my parents wanted to do was talk about marriage.

T: So what happened?

C: It was a bit difficult because although we are serious about each other we're not ready for marriage yet. We're still learning a lot about each other's culture.

T: Is it that different?

C: It's mainly interesting because there's so much to learn. He's far more open-minded than I am and has different opinions to me about nearly everything!

T: Do you argue a lot then?

C: No, James is too easy going and anyway I'm learning not to be so traditional!

习惯用语 1

① only interested in one thing: 只对钱感兴趣

T: 自从你跟 James 约会以来，遇到什么问题没有？

C: 有，我们经常在街上被路人盯着看，我真的很讨厌。但 James 却

Love 爱情

9

觉得无所谓，他早就习惯了。

T: 你为什么不喜欢？

C: 也许有些人会认为我是那种只关注钱的坏女孩，但 James 并不是
大款，他只是个英语老师，收入并不高。

T: 你父母怎么想？

C: 刚开始，我告诉我妈的时候，她连着两个礼拜都睡不着觉。后来
他们终于见面了，我父母想谈的却只有结婚问题。

T: 结果呢？

C: 有些困难，尽管我们彼此都是认真的，但我们还没有准备好要结
婚。对彼此的文化，我们还有很多需要学习的地方。

T: 两种文化有那么大的差别吗？

C: 文化差异才有趣呢，因为有很多要学的。他比我思想开放得多，
几乎对所有的事情，我们都有不同的见解。

T: 那你们经常发生争执吗？

C: 不，James 很随和，而且我也正学着不要如此传统。

Questions 1

1. Would you be interested in a cross-cultural relationship? Why? Why
   not?
2. What do you think about girls who go out with Westerners?

Dialogue 2

Gary: What do you like about Chinese girls?

Nick: My **present girlfriend**[①] is really lovely. She knows how to
look after me.

G: You are spoiled. How does she look after you then?

N: When I come home from work she makes me a cup of tea and
helps me to settle into my favourite chair and rearranges the
cushions.

G: Sounds perfect!

N: And then she cooks me a meal and makes sure I get the best of everything. She will take bits of food from the dishes and put it on my plate.

G: Aren't Western girls like that?

N: No! They're far too independent. They don't know how to look after a man like Chinese girls do. But they do like the way Chinese boys look after girls.

G: Can you introduce me to a Western girl?

N: Sure. I know a couple of Western girls who are looking for Chinese boyfriends.

G: But I've never had a Western girlfriend! I feel nervous. What should I do?

N: Treat her like a Chinese girl. Take her shopping, say nice things to her, buy her dinner and generally see that she gets everything she wants.

G: OK, I get the message. Why not call them now?

习惯用语 2

① present girlfriend: 现在的女朋友

G: 你喜欢中国女孩的什么?

N: 我现在的女朋友真的很可爱。她知道如何照顾我。

G: 你真是被惯坏了! 她是怎么照顾你的?

N: 当我下班回到家后, 她都会为我沏一杯茶, 然后扶我坐在舒适的椅子上, 还给我铺一个软软的垫子。

G: 听起来太完美了!

N: 然后她给我做晚饭, 保证让我吃得好, 吃饭的时候还总是给我夹菜。

G: 西方女孩不是这样吗?

N: 不! 她们相当独立! 她们不像中国女孩子那样知道如何照顾男人。但她们却很喜欢中国男孩照顾女孩的方式。

G: 你能给我介绍个西方女孩子吗?

Love 爱情

N: 可以啊，我认识几个西方女孩正想找个中国男朋友。

G: 可我从没交过西方的女朋友。我觉得有点儿紧张。我该怎么做？

N: 就把她当中国女孩子来对待，陪她逛街，赞美她，请她吃饭，尽量保证她得到所有她想要的。

G: 好的，我明白了。那现在就给她们打电话啊？

Questions 2

1. How does your girlfriend treat you?
2. How do you treat her?

Dialogue 3

Dan: How's your relationship going with Anna?

Mark: Not too good, to be honest.

D: What's the problem? I thought you two were **getting on like a house on fire**①.

M: We were to begin with but now we just seem to argue all the time. I think we want different things.

D: Such as?

M: Anna wants to improve her English whereas I want a lot of loving.

D: I see. So why don't you want to teach her English?

M: I spend all day teaching so I don't want to be doing it in the evenings as well. And Anna always wants me to go shopping with her when I'd rather stay at home and do some writing.

D: What about the loving bit?

M: I have a **high sex drive**② and it was fine to begin with but now she's **cooled off**③ and doesn't seem to want it very much.

D: I think you've got to compromise a bit. I'm sure if you taught her English a couple of nights a week and occasionally went shopping with her then her loving side would return.

M: I'm sure you're right. I'll take her out to dinner tonight and we'll try to talk things through.

习惯用语 3

① get on like a house on fire: 在一起相处得极其融洽
② high sex drive: 性欲强
③ cool off: 不再喜欢，失去热情

D: 你和 Anna 相处得怎么样？

M: 老实说，不是很好。

D: 怎么了？我还以为你们相处得很不错呢。

M: 我们是想好来着，但现在似乎总是在吵架。我觉得我们的追求不同。

D: 比如呢？

M: Anna 想要的是提高她的英语水平，然而我想要的是爱情。

D: 我明白了。那你为什么不想教她英语呢？

M: 我整天都在教课，所以我不想晚上还要教她。而且 Anna 总是想让我陪她去逛街，但我却想待在家里，写点儿东西。

D: 那你们俩那方面怎么样？

M: 我的性欲很强。刚开始我们在性方面还比较和谐，但现在她冷淡下来了，好像不是很想要性生活。

D: 我觉得你得学会退让。要是你能每周教她几次英语，偶尔再陪她逛逛街，我保证，她的性爱激情很快就会回来的。

M: 你说得没错。我今晚就约她出去吃饭，跟她好好谈谈。

Questions 3

1. What's the most important thing in a relationship?
2. What should you do if you both have different aims?

Love 爱情

Sex
3 性

‖ Background Information ‖

Sex is where two people（male and female）have sexual intercourse. According to the *Beijing Star Daily* 57% of men want to marry a virgin while only 35% of women hope their husband will be a virgin on their wedding night. A 2006 survey of more than 22,000 college students in Zhejiang showed that 13% have had sexual experience. For male students it's 18% and for females it's 9%. The students' first sexual experience came at 19.5 years of age on average.

A 2006 poll of 1,790 adults by the British newspaper *The Observer* found that the number of people having sex before 16 years old has fallen from 32 percent in 2002 to 20 percent in 2006. The age at which women first have sex is now 17.44 compared with 18.06 for men.

性，就是两个人（男和女）发生性关系。根据《北京星报》的调查显示，57%的男性希望娶一位处女，而只有35%的女性希望她们的丈夫在洞房之夜时仍为处子之身。2006年，一项对浙江省超过22000名大学生的调查显示，他们当中有13%的人有过性经验，其中，男生占18%，女生占9%，学生们第一次性经历的平均年龄是19.5岁。

2006年，英国《观察家》报一项对1790名成年人的调查显示，在16岁前便发生性行为的比例由2002年的32%，下降到了2006年的20%。女性第一次发生性行为的平均年龄为17.44，男性为18.06。

背景信息

Dialogue 1

Jill: Why is it, do you think, that Chinese women are so attracted to Western men?

Helen: Well, I have **a low opinion**^① of those women who go out with Western men.

J: Oh why?

H: I think some Chinese women think that Western men are richer than Chinese men. Or some women want to live abroad and that's why they date Western guys. There is no real love between them.

J: I don't agree. At least Simon and I are not like that.

H: Are you dating with a Western man now?

J: Yes, I am.

H: Can you tell me why you like him, if it's not because he is rich or he can take you to a Western country?

J: Well, first of all, I love his blue eyes. When he looks at me with his blue eyes, I feel faint.

H: Their blue eyes are different than ours, but can they really make you faint?

J: Yes, when you have love for each other and look at each other you will know.

H: Wow, what else? You can't marry someone just to look at each other every day, can you?

J: And secondly, he is very romantic. He holds my hand wherever we go. And he always acts like a gentleman.

H: Well, I think Chinese men also can do that.

J: And the last reason is the most important reason and that is we have very good sex feeling in bed. Which I never had with a Chinese man.

H: Aha, now I understand.

Love 爱情

15

J: 你说为什么中国女人会对西方男人有那么大的吸引力?

H: 我可不喜欢那些嫁给老外的女人。

J: 为什么?

H: 我觉得有些中国女人就是觉得老外比中国人有钱,或者是因为她们想到国外去定居,所以才和老外交往。他们之间并没有真爱。

J: 我不同意,至少我和 Simon 之间不是那样的。

H: 你在和老外谈恋爱?

J: 是的。

H: 那你能不能告诉我,假若不是因为他有钱,不是因为他能带你出国,你究竟喜欢他什么?

J: 嗯,首先,我喜欢他的蓝眼睛。当他用那双蓝色的眼睛注视着我的时候,我简直都要醉了。

H: 老外的蓝眼睛确实与我们不同,但这真的会令你如痴如醉吗?

J: 当然了,当你们之间产生爱意,彼此凝视的时候,你就会明白了。

H: 哇噢,别的呢? 你和一个人结婚总不能就是为了整天两个人对着看吧?

J: 其次,他非常浪漫。无论走到哪里,他都拉着我的手,像个绅士。

H: 我觉得中国男人也可以做到啊。

J: 最后,也是最重要的原因,我们的性生活很美满。这是我以前与中国男人从未有过的。

H: 哈,我明白了。

Questions 1

1. Why do you think Chinese girls date Western men?
2. Do you think Western men are better lovers than Chinese men?

Dialogue 2

Tony: Can you tell me whether a western guy would like to choose a girl with less sex experience or more sex experience as his partner?

Mark: I think it depends on different people, but for me and most of my friends, we'd like to choose a girl who is sexually experienced. How about Chinese men?

T: I can't speak for all Chinese men but I prefer a girl with no sex experience.

M: I understand. In Chinese tradition, a girl should be a virgin before marriage. But for us, if a girl is still a virgin before her wedding night, we will wonder about whether we will be sexually compatible.

T: Is sex really so important to you western men?

M: Yes, it is the most important thing in a relationship. If the sex is good, every big problem will become a small problem. If the sex is not good, every small problem will become a big problem. So both husband and wife should be **good in bed**①.

T: I see. In China, if a woman is sexually experienced, we will think that she had many boyfriends in the past and may be very easy to sleep with. And if we marry this kind of woman, she might be unfaithful to her husband in the future. So we will still want to choose an innocent faithful girl to be with.

M: But what about if you can't be sexually satisfied with her?

T: I can teach her. And I think a peaceful life is more important than crazy sex in a relationship.

M: Then, I hope you won't have sex out of marriage and be patient to teach your wife. What's more a woman's sex experience cannot only come from her man but also come from her study and self awareness. Yoga and the *Kama Sutra* are good for women to study. On the other hand, even if she had many boyfriends in the past, it doesn't mean she will not be faithful to her husband in the future.

T: Maybe, you are right. But still an innocent girl appeals to me more.

M: I am not trying to convince you. And I do hope you find your ideal girl.

习惯用语 2

① good in bed: 性技巧高超

T: 你能不能告诉我，西方男人谈对象时，是愿意找个性经验多的还是少的女孩？

M: 我觉得这得根据各人而定，不过对于我和我的很多朋友来说，我们更愿意选择那些性经验很丰富的女孩。中国男人会怎么选择？

T: 我不能说全部，但至少我更喜欢没有性经验的。

M: 我明白。在中国的传统里，女孩在婚前应该保持处女之身。但对我们来说，要是一个女孩直到她结婚之夜还是处女的话，我们就得考虑一下婚后的性生活是否会和谐了。

T: 性对于你们西方男人来说真的那么重要吗？

M: 是的，这是两人关系中最重要的部分。要是性生活美满，所有的大问题都会成为小问题。反之，所有的小问题都会成为大问题。所以，夫妻二人都应当是"床上高手"。

T: 我明白了。在中国，要是一个女人性经验很丰富，我们就会觉得她之前有过很多男朋友，而且很容易就被搞上床。要是和这种女人结了婚，将来她也许会对她的丈夫不忠。所以，恋爱时，我们还是愿意选择那些单纯、忠诚的女孩。

M: 那要是你和她无法在性方面得到满足怎么办？

T: 我可以教她。我觉得两个人相处，和谐平静的生活比疯狂的性爱更重要。

M: 那我希望你以后不要出轨，耐心地教好你的妻子。另外，女人的性经验，不仅仅来自她的男人，还要靠她自身的学习与领悟。比如女人练习瑜伽，以及看看《爱经》（印度8世纪时一部关于性爱和性爱技巧的著作——译者注）就很不错。另外，即使她过去有过很多男朋友，也不能说明她将来就对她的丈夫不忠啊。

T: 也许你是对的。但是，纯真的女孩对我还是更有吸引力的。

M: 我也不想试图说服你，只是希望你能够找到自己理想中的女孩。

Questions 2

1. Do you think virginity is important?
2. Would you like to marry a virgin?

Dialogue 3

Nick: What are Chinese traditional views on sex?

Clint: According to *The Book of Rites* men should not marry until the age of thirty and women at the age of twenty.

N: What's the principle behind that?

C: It's all to do with health. If a man has too much sex then it consumes one's essence of life stored in the kidneys and this is one of the main causes of **premature senility**[1] and declining health.

N: How much sex should a man have then?

C: Wang Gui, who was a scholar in the Yuan dynasty, said that it was best for people at the age of thirty to have sex once every eight days and those aged forty once every sixteen days.

N: What about those over that age?

C: Men aged fifty should have sex once every twenty days and those aged sixty and over not at all but if they can't restrain themselves it should only be once a month.

N: So restricting sex means that men can live longer, right?

C: Right.

N: I think in the West we would rather have a short life with plenty of sex than a long life without it!

C: Well, it's your choice.

Love 爱情

19

① premature senility: 未老先衰

N: 对于性，中国的传统观点是什么？

C: 根据《礼记》记载，男子要到 30 岁才能结婚，女子要到 20 岁。

N: 这背后的道理是什么？

C: 全都与人体健康有关。要是一个男人纵欲过度，就会消耗肾脏所贮藏的、人体必要的精气，这是未老先衰、精力衰退的主要原因。

N: 男人的性生活频率应该怎样才算合适呢？

C: 根据中国元朝学者王魁的观点，对于 30 岁的人而言，每 8 天一次性生活，对于 40 岁的人来说，每 16 天一次。

N: 要是超过了那些年龄呢？

C: 50 岁的男人，每 20 天一次。60 岁以及年龄更大的人，就不该有性生活了。如果无法抑制的话，也应该是一个月一次。

N: 所以，禁欲就意味着男人可以更长寿，对吗？

C: 没错。

N: 我觉得在西方国家，我们宁愿短命也要有丰富的性生活，而不愿意为了长寿而不过性生活。

C: 那是你的选择。

Questions 3

1. What's your opinion about Chinese traditional views on sex?
2. Would you like a short life with plenty of sex or a long life with no sex? Why?

4 婚 姻

Marriage

Background Information

Age when women get married

Country	1970	2003
U.S.	20.8	25.3
U.K.	22.5	28.3
China	20.2	24.5
Japan	24.5	27.8

In China women are under huge pressure from their parents to marry before the age of 30, men before 35. In the U.K., it is common for people to live together before marriage.

女子结婚年龄

国家	1970	2003
美国	20.8	25.3
英国	22.5	28.3
中国	20.2	24.5
日本	24.5	27.8

在中国，要是女人到了30岁，男人到了35岁还不结婚的话，就会面临来自父母的巨大压力。在英国，未婚同居是非常普遍的事情。

背景信息

Dialogue 1

Cindy: If you can choose, will you marry a western man or a Chinese?

Lily: Why? Are you having to make a choice now?

C: Yes, it's really giving me a headache. I feel I am **between a rock and a hard place**①.

L: Let me guess, you have a western boyfriend but your parents want you to marry a Chinese, right?

C: You are smart. I got to know a western boyfriend; he is very romantic with a great sense of humour. But my parents said Chinese should marry Chinese.

L: I had the same problem like you before.

C: Then how did you deal with it?

L: I chose to continue dating with my western boyfriend. But we broke up because we were too different.

C: Oh, that's a pity, what was the reason you broke up?

L: I think western men are romantic and make better lovers. But when it comes to marriage, there are many culture shocks.

C: So what kind of culture shock did you come across?

L: First of all, he asked me to sign a **pre-nuptial contract** \ ②. Which said we should pay our own bills after marriage, we manage each other's money by ourselves, and that in the event of divorce I would not be entitled to half his money but only a small amount.

C: Oh, really? That's more like doing business not a marriage.

L: Yes, he said he hoped to live in their western way even with a Chinese girl.

C: So, it's difficult to get along with a western guy?

L: Not really, you should find one who can accept the Chinese way of life or you should accept their western way of life.

① between a rock and a hard place: 左右为难，进退两难
② pre-nuptial contract: 婚前协议

C: 要是可以选择的话，你会嫁给老外还是中国人？

L: 怎么问这个？你现在正面临这种选择吗？

C: 是啊，真是让人头疼。我觉得自己是进退两难。

L: 让我猜一下，你交了个外国男朋友，可你的父母想让你嫁个中国人，对吗？

C: 你太聪明了！我认识了一个外国男孩，他很浪漫而且幽默。可我父母说，中国人就该嫁给中国人。

L: 我原来也遇到过类似的麻烦。

C: 那你是怎么解决的？

L: 我选择了继续与外国男友约会。不过最终我们还是分手了，因为我们的差异太大。

C: 噢！真遗憾！你们为什么分手？

L: 我觉得西方男人很浪漫，是很好的恋爱对象。但谈到婚姻，就有很多文化冲击。

C: 比如呢？你遇到了什么冲击？

L: 首先，他要我签一份婚前协议，要求我们婚后各自管理自己的钱财，承担自己的花销。万一离婚了，我也不能得到财产的一半，而只是一小部分。

C: 真的？我觉得这更像是在做买卖，而不是在谈婚论嫁。

L: 没错，他说即使是娶了一个中国女孩，他也想按照他们西方的方式来生活。

C: 所以，跟西方男人相处很困难？

L: 也不是，你得找一个能够接受中国生活方式的人，或者你就接受西方的生活方式。

Questions 1

1. Would you like to marry a Chinese or a foreigner?
2. Would you like to change to a western way of life?

Love 爱情

Dialogue 2

Wendy: How is your marriage going so far?

Cathy: Well, not so well.

W: Oh, why?

C: It's because of my mother-in-law. She is a very traditional Chinese woman.

W: Well, I think in-laws always cause problems no matter in what culture.

C: You know she wants me to have a baby with my husband right away.

W: I see, she is eager to see her grandchildren.

C: But when and whether to have a baby is absolutely our own business. And you know I just got a good position in our company. If I have a baby now, I will lose a good career opportunity.

W: I understand you well. I think the older generation have different concepts than us. They think women should **put family first**①.

C: I have tried my best to get along with her, and it seems no matter what I do, she will not be happy with me. You know we live in the same house and I really want to set up a good relationship with her.

W: Don't worry, it takes time to get used to a new environment. In fact, I also had problems with my mother-in-law when my husband and I lived in the U.K.

C: What was your problem then?

W: Well, my mother-in-law is an English lady; she retired several years ago. My husband and I are both very busy with our work every day. So I asked her to come to our house and look after our baby sometimes. Which she did but she interfered in everything and complained about our life style too much. I know she loves us but I don't want her to control our lives.

C: I guess that is your western concept too.

W: Yes, mothers will always act like mothers and always regard us as children no matter how old we are!

习惯用语 2

① put family first: 把家庭放在第一位

W: 婚后的生活如何?

C: 嗯, 不是太好。

W: 噢? 怎么了?

C: 因为我婆婆, 她是一个非常传统的中国女人。

W: 我觉得无论是在什么文化背景之下, 婆媳之间总会出现一些问题。

C: 你知道嘛, 她希望我和我老公赶紧生个小孩儿。

W: 我明白了, 她是想抱孙子了。

C: 但是, 是否要孩子, 什么时候要孩子, 完全是我们自己的事情。你知道, 我最近刚刚升职, 要是现在要小孩儿的话, 我就会失去一个极好的发展事业的机会。

W: 我非常理解你。我觉得我们与长辈之间的观念有所不同。他们觉得女人就应该把家庭放在第一位。

C: 我已经尽力与她和睦相处了, 但似乎不论我做什么, 她都不会开心。你知道, 我们住在一起, 我真的希望能跟她相处得融洽些。

W: 别着急。适应新环境需要时间。事实上, 我和我老公住在英国的时候, 也跟我婆婆出现过类似的问题。

C: 你遇到的是什么麻烦?

W: 我婆婆是一位英国女士, 几年前就退休了。我和我老公每天的工作都非常繁忙。所以我就请她来我们家, 有空的时候帮忙照顾一下小孩儿。尽管她这样做了, 但她什么事都要干涉, 而且总是抱怨我们的生活方式。我知道她很爱我们, 但我并不想让她控制我们的生活。

C: 我估计那也是你的西方观点。

W: 是的，母亲永远都是母亲。不论我们多大了，她始终把我们当作孩子。

Questions 2

1. What problems do you think in-laws will cause?
2. How would you solve those problems?

Dialogue 3

Barrymore: What age can you marry in the U.K.?

Nick: Legally it's eighteen but you can get married at sixteen if you have your **parent's consent**①.

B: That's very young. What do you do if you can't get their consent?

N: Some couples in England used to elope across the border with Scotland and got married in a blacksmith's forge in a small town called Gretna Green.

B: That sounds very romantic. So did that make it legal?

N: Yes. What about here in China?

B: Men can get married at twenty-two while for women it's twenty.

N: I know it used to be illegal for students to get married while at university but I gather they can do now.

B: That's right and I even heard of one university allowing **maternity leave**②.

N: In England mothers get one year's maternity leave from work and I believe that there may be legislation soon allowing fathers six months paternity leave.

B: Wow! Why?

N: I think it's so that fathers can bond with their child and help to keep the marriage secure.

① parent's consent: 父母的同意　　② maternity leave: 产假

B: 在英国，多大年龄才能结婚?

N: 法律规定是 18 岁，但要是你父母同意的话，你也可以 16 岁就结婚。

B: 那太小了。要是不能得到父母的同意怎么办?

N: 有些英格兰的情侣会跑到与苏格兰接壤的地方，在一个叫做 Gretna Green 的小镇上的一个铁匠铺完婚。

B: 听起来很浪漫啊! 这样做合法吗?

N: 合法。中国的情况是怎样的?

B: 男子 22 岁，女子 20 岁，可以结婚。

N: 我知道过去大学生结婚是违法的，但现在可以了。

B: 没错。我还听说有一所大学允许休产假呢。

N: 在英国，母亲有一年的产假。我相信很快就会立法，允许父亲们休 6 个月的产假。

B: 哇! 为什么呀?

N: 我觉得通过这样，父亲就可以使自己的亲子关系和婚姻关系更安全和稳固。

Questions 3

1. Do you think eighteen is too young to get married? Why?
2. Do you think all universities should allow maternity leave?

Love 爱情

5 考入大学

Getting to University

⫿ **Background Information** ⫿

The college entrance exam was reinstated in December 1977 by Deng Xiaoping after a ten year gap. According to the Ministry of Education, some 36 million students have been admitted to tertiary institutions such as universities and colleges since then. In 2006 9.52 million people applied to take the exam of which 8.8 million took the test with 5.4 million being enrolled. Over 9.5 million students sat the exam in 2007 with 5.67 million succeeding.

　　1977 年 12 月，邓小平同志恢复了停滞十年的高考制度。根据教育部的统计，从那时起，约有 3600 万人通过参加高考进入了高等院校学习。2006 年，952 万人报名考试，实际有 880 万人参加了考试，其中 540 万人被录取进入了高等院校。2007 年，超过 950 万人报名参加高考，从中录取 567 万人。

背景信息

Dialogue 1

Ann: Hi Nick, could you tell me something about British culture?

Nick: Sure. Where should I begin? What are you specifically interested in?

A: Well as I am a student tell me something about British students.

N: You have more universities in Beijing than we do in the whole of the U. K. But whereas about 7% of Chinese students go to university the figure in the U. K. is 30%.

A: That's because our population is so large. What's yours?

N: Around 60 million. So really in the U. K. any student who wants to can go to university. They can choose as many as six universities and there are many majors to choose from.

A: Do they have to pass a common entrance exam like we have to?

N: No. We take what are known as "A" levels in several subjects. When we are sixteen we start studying these "A" levels for two years.

A: How many "A" levels do you need to get into university?

N: Two, but the top 0**unis**[①] would prefer three. It's common to choose a university major based on one of your "A" levels. So if you got a good geography "A" level then you might go on to study geography at uni.

A: I heard something about gap years. Can you tell me something about that?

N: About 7. 5 percent of students applying to university defer their university offer for a year so that they can do voluntary work overseas. They often find that a break refreshes them and also helps them become more mature.

A: Sounds like a good idea but I don't think it will ever **catch on**[②] in China!

① uni: 大学（简称）　　　　② catch on: 流行起来，被（人）接受

A: 嗨 Nick，跟我谈谈英国的文化好吗？

N: 好啊。从哪儿开始说起呢？你对什么特别感兴趣？

A: 嗯，既然我是个学生，那就跟我说说英国的学生吧。

N: 北京的大学数量比整个英国的都多。但却只有 7% 的中国学生可以上大学，在英国，这一比例是 30%。

A: 那是因为我们的人口基数大。英国有多少人？

N: 大概 6 千万人。所以，对于英国的学生来说，只要他想上大学，就可以上。他们可以在 6 所大学中选择，还可以有很多不同的专业供他们选择。

A: 他们要参加统一的入学考试吗？就像我们的高考一样？

N: 不用，我们学的课程叫 A-level，有好几门课，从 16 岁以后就可以学了，得学两年。

A: 要想进入大学得学几门 A-level 课程？

N: 两门，不过有些好大学更倾向于三门。通常你根据一门 A-level 课程来选择自己的专业。比如你的 A-level 课程中地理学得比较好，那你可能就会在大学里继续学习地理专业。

A: 我还听说你们有"休学年"，这是怎么回事啊？

N: 大约有 7.5% 的学生申请到了大学，然后可以延期一年入学，这样他们可以去海外做志愿者。他们经常会发现，这种休学不但可以增长见识，还可以让他们更加成熟。

A: 听起来不错，但要在中国实行恐怕不太可能。

Questions 1

1. Was it difficult for you to enter university? Why?
2. Would you ever consider doing a gap year?

Dialogue 2

Nick: So how do Chinese students get into university?

Ann: We have to pass the National Common Entrance Exam.

N: Is it very difficult to pass?

A: Depends on where you live. Cities like Beijing and Shanghai have their own version of the exam and you don't have to get as high a score as some other places.

N: Oh, why is that?

A: Someone said that it's to make it easier for government officials' children to get in but I don't know if that's true or not.

N: What if you live in the countryside?

A: Some provinces have lower scores so it's easier to get in and places like Tibet get **preferential treatment**[①].

N: I think that's fair because they don't have many universities there. Any other ways of getting to university.

A: If you are recommended by your high school then you don't have to take the exam.

N: That's good if you are an excellent student! Any others?

A: I heard that if your parents make a donation of about 500,000 yuan to a top university then they will let your children in. Does that happen in England too?

N: I've heard of some cases but I think they still have to pass exams and meet the university's requirements.

习惯用语 2

① preferential treatment: 优先待遇

N: 中国学生如何进入大学？

A: 我们得参加全国统一的入学考试。

N: 很难通过吗？

A: 那得看你生活在哪里。像北京、上海这种大城市，有自己的命题，考生不用像其他地方那样需要考很高的分数。

N: 为什么呢？

A: 有人说这样做是为了让政府官员的孩子更容易进入大学，但我不知道这是不是真的。

N: 那要是住在农村怎么办？

A: 有些省份的分数就比较低，进入大学也比较容易。比如西藏，就能有优先待遇。

N: 我觉得这很公平，因为那里没有那么多大学。还有什么别的方法可以上大学吗？

A: 要是你能得到高中的推荐，就不用参加高考。

N: 如果你是个优秀的学生，这方法还不错。别的呢？

A: 我听说，要是你的父母给大学捐款 50 万，你就可以进入这所大学。这种事情在英国会发生吗？

N: 我倒是听说过，不过我觉得一般学生还是要参加考试的，要达到大学的基本要求，才能上大学。

Questions 2

1. How easy or difficult was it for you to enter university?
2. How would you change the college entrance exam?

Dialogue 3

Nick: I was having breakfast this morning in a hotel and saw several mothers with their son or daughter. I guessed they are taking the Common Entrance Exam.

Ann: It's quite common. Parents want their child to be near the test centre and also that they might have a quiet place to rest and study.

N: There's huge pressure on children and parents to do well in the exam, isn't there?

A: Yes, I know some parents hire tutors for their children and also many go to lectures given by teachers, successful entrants and even parents.

N: Parents! What do you mean by that?

A: I heard that one parent whose children got into Beida, Tsinghua and Renda had joined the **lecture circuit**[①]. Parents will do anything to ensure that their child gets into university.

N: I was reading online that in Guangzhou some parents even block the traffic in some streets so that their kids will get a good night's sleep!

A: Well it is important that they pass and get into a good university. I even know that some students who get a high mark don't go to university that year but retake it the next year hoping to get into Peking or Tsinghua University.

N: Having taught at both of them I know that many students feel disappointed when they get there.

A: Oh, why is that?

N: They were the best students in their province but suddenly find that they are no longer the best and that maybe they are not as smart as they thought they were. They've achieved their dream but realise there is still a long way to go.

A: Life's like a box of chocolates. You never know what you're going to get!

习惯用语 3

① lecture circuit: 巡回演讲

N: 今天早上我在饭店里吃早餐时，看到几位母亲带着自己的儿女。我想他们应该是去参加高考的。

A: 这太常见了。父母想让孩子离考场近一点，同时也能有个安静的地方来休息和复习。

N: 要想考好，家长和孩子都要承受巨大的压力，对吧？

A: 没错。我知道很多家长都给孩子请家教，还参加各种讲座，比如老师辅导，或者是高考优胜者经验交流，甚至还有家长座谈。

N: 家长座谈！什么意思？

A: 我听说，那些考进北大、清华和人大的学生家长，组成了一个高考巡回演讲团，与考生家长交流经验。为了让孩子进入大学，家长可谓是煞费苦心，想尽办法。

N: 我在网上看到文章说，在广州，为了能让孩子夜里休息好，有些家长甚至去封锁附近的交通。

A: 是啊，通过高考，进入理想的大学，这可是大事儿！我甚至知道有些学生当年已经考了很高的分数却还要复读，为的就是进入北大或者清华。

N: 我在这两所大学都教过课，我知道很多学生进入了清华或北大却很失望。

A: 为什么？

N: 他们都是省里的尖子生，进入大学后忽然发现自己并不那么优秀了，或者觉得并不像自己想象中的那么聪明。他们完成了自己的梦想却发现其实还有很长的路要走。

A: 生活就像一盒巧克力，你永远不知道将会得到什么。

Questions 3

1. What special preparations did you make for the exam?
2. How do you think you compare with other students?

6 校园生活 1

Campus Life 1

‖ Background Information ‖

Campuses can range from the very small (e.g. Beijing University of Posts and Telecommunications) to the very large (e.g. Tsinghua). The main gate into the campus is usually the East Gate. On campus there would be the usual facilities such as teaching buildings, dormitories, library, sports ground, dining rooms, supermarket, etc. Most students would own a bike but often these are stolen. Dining rooms are cheap and crowded at meal times but the food is not of a high quality. Around 5.3 million students (out of 9.5 million) were admitted to universities in 2006 which was an increase of more than 10 percent from 2005. According to a 2006 survey by the Ministry of Education about 83 percent of college students are short-sighted.

大学校园的面积可能很小（比如北京邮电大学），也可能很大（比如清华大学）。大学的主校门通常都是东门。校园内都是些寻常的建筑，比如教学楼、宿舍楼、图书馆、运动场、食堂、超市等等。很多学生都有自行车，但却经常被盗。食堂的饭菜价格并不高，当然，质量也不怎么样，用餐高峰的时候还会很拥挤。2006 年有大约 530 万学生进入大学（950 万人参加考试），比 2005 年增加了 10%。2006 年教育部的调查显示，83% 的大学生都患有近视。

背景信息

Dialogue 1

Ann: What do students do in their spare time?

Nick: There is a strong drinking culture in Britain so most students would socialize in pubs. Sometimes they drink too much.

A: If Chinese students drink they mainly do it at mealtimes.

N: I think drinking to excess is a bad habit. When I was in Cambridge many of the colleges had drinking clubs and students would often get drunk and then **streak**① across the college lawn.

A: I've heard something about those college lawns. Aren't they reserved for **dons**② only?

N: That's right. I think originally it was a privilege given to them to make up for the fact that they had to be unmarried.

A: Many Chinese students like karaoke. What about British students?

N: I once read that 70% of Chinese students go to a karaoke bar but in Britain it would be very few, perhaps only 5%.

A: Do British students study hard?

N: Perhaps only just before exams! I find it strange here that you always see students studying in the library and classrooms but in the U. K. some students would study in the library but never in classrooms.

A: Why is that?

N: Our libraries are big enough and we don't have as many students as you do.

习惯用语 1

① streak: （在公共场所）裸体飞跑
② don: 牛津或剑桥大学教授

A: 英国学生在课余时间都做些什么?

N: 英国人都很爱喝酒，所以大多数学生都会去酒吧，有时候还会酗酒。

A: 中国学生也喝酒，但大多在吃饭的时候才喝。

N: 我觉得饮酒过量是个坏习惯。我在剑桥读书的时候，很多学院都有喝酒俱乐部，学生们经常喝醉，还在学院的草地上裸奔。

A: 我听说过学院草坪的事情，那不是只给导师们留着的吗？

N: 没错。我想主要是因为当时剑桥的老师们必须是未婚，所以才有拥有草坪这样的特权，以作为补偿。

A: 很多中国学生都喜欢唱卡拉 OK。英国学生呢？

N: 我看到过报道说 70% 的中国学生都去 KTV，可在英国，这一比例很小，大概只有 5%。

A: 英国学生学习刻苦吗？

N: 可能只有在考试前会刻苦一点儿。可在这里总是能看到学生们在图书馆或教室上自习。但在英国，学生们会去图书馆，从不去教室。

A: 为什么？

N: 我们的图书馆非常大，而且学生也不像你们这儿那么多。

Questions 1

1. Do you ever drink too much?
2. Have you ever done something stupid when drunk?

Dialogue 2

Ann: What about living conditions?

Nick: Students can choose one of three options. They can live in university accommodation, rent privately or live at home.

A: We are not allowed to live off campus as the university is responsible for our safety. What is university accommodation like?

N: We usually have single study bedrooms and maybe share a kitchen and shower with about ten other students.

A: As you know we have dormitory-style accommodation here in China so there could be anything between four and eight students in a dorm.

N: Another thing is that we can come and go at whatever time we

Education 教育

like unlike here where you have to be in your dorm by eleven.

A: That's right. So if we get locked out we have to go to an Internet bar to spend the night.

N: We also have hot water 24/7[①] unlike here where you have to fill up your vacuum flasks every day!

A: And we have to go to the bath house several times a week! And we don't have private showers either!

N: I know that some Chinese dormitories have showers and toilets but I suppose they're quite expensive.

A: That's right. I can't wait to study in Britain!

习惯用语 2

① 24/7: 每天 24 小时，每周 7 天

A: 英国学生的住宿条件怎么样？

N: 学生有三种选择：住宿舍，自己租房子或者住在家里。

A: 学校要对我们的安全负责，所以不允许我们住在校外。大学宿舍的条件如何？

N: 通常都有单独的一间卧室兼作学习之用，但要与大约 10 名学生共用厨房和浴室。

A: 你知道，在中国就是住集体宿舍，要 4 到 8 个人共住一屋。

N: 还有一点就是我们可以随时进出宿舍，不像这里，11 点以前一定要回宿舍。

A: 没错。要是我们被锁在了外面，就只能在网吧过夜了。

N: 我们还有 24 小时的热水。在这里却每天都要去打热水。

A: 每周还得去几次浴室，我们没有私人的浴室。

N: 我知道有些中国学生的宿舍就有浴室和卫生间，但我想那一定很贵。

A: 没错。我都想去英国读书了！

Questions 2

1. What are the main differences between British and Chinese students?
2. Would you like to study in a British university?

Dialogue 3

Lucy: What are some of the odd things you've noticed about campus life?

Nick: Sometimes I see female students going to the bath house and they're wearing their pajamas! We would never be seen outside the house wearing our night things!

L: Maybe they are getting ready for bed!

N: Another thing is that students always carry vacuum flasks to fill up with hot water. We have hot water in our rooms so that's strange too. What about you?

L: I studied at Cambridge for a year and I would see lots of strangely dressed students going for a **pub crawl**[①] in the evenings.

N: Yes, I saw them too. Some were even dressed in Roman togas!

L: British students seem to spend a lot of their time in pubs. We spend a lot of time in the library or in classrooms studying.

N: Or playing sports! I see so many students playing basketball or badminton in the evening even when it gets too dark to see!

L: Sports are very popular here in China but in Britain you don't really do that much sport.

N: Another odd thing I see would be girls holding hands. If two girls their age did that in Britain we would think they were gay!

L: It's just because girls get very close to each other and treat them as sisters.

N: I think that's very touching but I'm glad the boys don't do that!

① pub crawl: 在一个晚上连续光顾好几个酒吧

L: 在中国，你有没有发现校园生活里的一些奇怪事情？

N: 有时候我看到女生穿着睡衣去浴室。我们从来不会在房间外面穿睡衣的。

L: 大概她们是准备睡觉了。

N: 还有一点就是学生们总是拿着水壶去打热水，这很奇怪，我们在房间里就有热水了。你呢？

L: 我在剑桥学习过一年，总是看到很多奇装异服的学生外出去酒吧。

N: 是的，我也能看到。有的甚至穿那种罗马人的托加袍。

L: 英国学生似乎会把很多时间都花在酒吧里。我们却把时间都花在图书馆和教室里。

N: 还有做运动。晚上我总能看到学生在打篮球和羽毛球，即使天已经黑得看不见了还在打。

L: 运动在中国很流行，但在英国却不是这样。

N: 还有一件奇怪的事情就是女孩子会手拉着手。在英国，如果两个女孩手拉着手我们会觉得她们是同性恋。

L: 这不过是因为女孩子们感情很好，把彼此当作姐妹罢了。

N: 太吓人了，好在男孩子不会这么做。

Questions 3

1. What are some of the odd things you have seen on campus?
2. Do you ever hold hands with a girl? Why?

7 校园生活 2

Campus Life 2

‖ **Background Information** ‖

According to the National Bureau of Statistics China spending on education as a percentage of GDP was

China 5. 29

U. K. 6. 1

The university yearly tuition as percentage per capita of GDP

China 37

U. K. 12

In 2005, China spent 842 million yuan or 2. 82 percent of its GDP on education, much lower than the 5. 2 percent international standard.

国家统计局的调查显示，教育开销占 GDP 的比例为：

中国 5. 29%

英国 6. 1%

大学学费支出占人均 GDP 的比例为：

中国 37%

英国 12%

2005 年，中国将 8.42 亿元人民币用在教育上，相当于 GDP 的 2.82%，比国际标准的 5.2% 低了很多。

背景信息

Dialogue 1

Nick: Chinese students always seem to drink lots of **bottled water**①. Why is that?

Paula: I think it's because **tap water**② is not safe to drink unless you boil it first.

N: In Britain tap water is safe to drink. I would often drink a glass of water straight from the tap.

P: I think that most of our water would be safe to drink but it's the pipes that carry them that are old and may contain chemicals that are harmful to water.

N: I once read that tap water in London has been recirculated at least seven times but it's still safe to drink because it's been treated.

P: Recirculated! What does that mean?

N: It means that it's been drunk, gone through the human body, been treated and then passed into the water supply system again.

P: Urrgh! I wouldn't like to drink that!

N: It's perfectly safe I assure you. But I know that more and more people like to drink mineral water.

P: What's that?

N: Water from springs that may contain small amounts of minerals. It's supposed to be healthy for you.

P: I'll drink to that!

习惯用语 1

① bottled water: 瓶装水　　　② tap water: 自来水

N: 中国学生总是喝很多瓶装水，为什么？

P: 因为自来水喝起来不安全，除非先把它煮开了。

N: 在英国，自来水就是可以安全饮用的。我就经常直接用杯子接来喝。

P: 我想我们的大部分水质也是安全的，只是运输的管子已经老化了，可能含有污染水质的化学物质。

N: 我曾经看到过报道，称伦敦的自来水经过了至少七次再循环却仍然是安全的，因为经过了处理。

P: 再循环？什么意思？

N: 就是说，被人喝了，然后又排出体外，经过处理后继续进入水供应系统。

P: 啊?! 我可不想喝这种水！

N: 那是非常安全的，我保证。但是我知道越来越多的人喜欢喝矿泉水。

P: 那是什么？

N: 就是从泉里流出的水，会含有少量的矿物质。人们都觉得这有益健康。

P: 那我也喝！

Questions 1

1. Do you drink a lot of water? Why?
2. Would you drink British tap water?

Dialogue 2

Elaine: Is this a typical British pub, then?

Nick: Well apart from the Chinese staff! The Goose and Duck is really well known in Beijing.

E: It's certainly **packed**① in here! I can hardly hear myself think!

N: It's really big inside. It's also known as a sports bar. Do you see that large screen over there? They show live sports matches.

E: And they've got lots of TV's tuned in to foreign channels too.

N: You can also play pool here too. Do you fancy a game?

E: Maybe later. I want to have a look around first.

N: Do you want to have a drink? There are plenty of British beers to choose from.

E: What do you suggest?

N: You could try a bottle of Newcastle Brown which is a kind of brown ale or there's Guinness which is black stout.

E: Are they both strong?

N: A bit yes so maybe a Carlsberg?

E: OK and then we could have a game of darts.

E: 这是那个典型的英国酒吧吗?

N: 差不多吧, 除了是中国雇员。"鹅和鸭"酒吧在北京很有名。

E: 也很拥挤! 吵的我都听不见自己说话了。

N: 这里面很大。这也是个运动吧。看到那个大屏幕了吗? 那里会播放现场转播的体育比赛。

E: 还能看到很多外国频道。

N: 你还可以在这里打台球。想玩儿吗?

E: 待会儿吧。我想先四处看看。

N: 想喝点儿什么? 这儿有很多英国啤酒可供选择。

E: 你有什么建议?

N: 你可以尝一瓶 Newcastle Brown, 是一种棕色麦芽酒。或者 Guinness 黑啤酒, 这种酒比较烈性。

E: 这两种酒都很烈吗?

N: 有点儿, 要不你来一瓶嘉士伯?

E: 好的。然后咱们玩儿一盘飞镖。

Questions 2

1. Do you like going to pubs? Why? Why not?
2. What do you like to do in a pub besides drinking?

Dialogue 3

Nick: I really hate mosquitoes!

Sunny: Don't we all! They've been bothering you again, have they?

N: You bet they have. We don't have them in Britain. I keep my windows closed but they still keep getting in and then I wake up in the middle of the night covered in bumps and hurting all over. I then spend hours hunting them down and killing them with a book.

S: Didn't know mosquitoes could read! Is the book one of yours? Did they die of boredom?

N: Ha ha, very funny. You know, it's strange when I kill one and there's a little red smear on the wall and I realize that's my blood!

S: You should do what we do.

N: Oh, what's that, then?

S: Burn some mosquito coils. That will frighten them off.

N: I've tried that but it doesn't seem to work for me.

S: You could get one of those electric things. They do the same thing.

N: I've got one but it doesn't work either.

S: You could get a bottle of mosquito repellent from the shop. Just put some on your face and other **exposed parts of your body**[①] and you'll be OK.

N: My girlfriend is going to love that!

习惯用语3

① exposed parts of sb.'s body: 暴露在外面的身体部分

N: 我真的很讨厌蚊子!

S: 我也是!它们又来搔扰你了,是不是?

N: 是啊!在英国就没有蚊子!我已经把窗户都关上了,可还是有蚊子进来,我半夜醒来的时候满身都是包,浑身痒痒,然后再花好几个小时用书把蚊子轰下来然后打死!

S: 我还真不知道蚊子还能读书!是你写的书吗?它们是不是因为读你的书郁闷而死?

N: 哈哈!真搞笑!你知道嘛,我打死一只蚊子,在墙上留下一个小红点儿,想到这就是我的血时,那种感觉真的好奇怪。

S: 你应该跟我学。

N: 什么?

S: 点蚊香啊。这样就可以把蚊子熏走了。

N: 我试了,可是似乎对我不管用啊。

S: 那你就买个电的,效果一样。

N: 买了,也没用。

S: 那就去商店买一瓶驱蚊剂。把你的脸,还有暴露在外面的身体部分都涂上,这样就行了。

N: 我女朋友肯定喜欢这个!

Questions 3

1. How do you deal with mosquitoes?
2. What other ways besides those mentioned above can keep mosquitoes away?

8 课程和老师

Classes and Teachers

‖ Background Information ‖

Chinese students usually have a lot of classes which start at 8 am and might finish in the evening. Classrooms normally have fixed desks and there would be a blackboard but there is a trend to update classrooms to include multi media. Often the start and end of a class begins with a bell. The usual size of a class would be 30 but it could be as low as 20 or as high as 40. Classes are really double periods of 45 or 50 minutes each with a break of 10 minutes in between. The academic year consisting of two semesters starts in September and finishes end of June or early July.

中国学生的课程很多，从早上 8 点也许能一直上到晚上。教室的桌椅都是固定的，还有黑板。不过，最近流行给教室安装多媒体设备。上下课都有铃声，一个班通常会有 30 人，也有可能 20 多或者 40 多人。每节课有 45 或 50 分钟，两节课中间有 10 分钟的休息时间。一学年有两个学期，九月开始，六月下旬或者七月上旬结束。

背景信息

Dialogue 1

Nick: You've had both Chinese and Western teachers. Can you tell me the difference between them?

Maggie: I think that Chinese teachers are stricter and have no sense of humour. They would never tell jokes like you do.

N: Perhaps they don't know any! I like to make my students **feel at ease**[①]. Anything else?

M: They always give us plenty of homework.

N: So I give you too little. Well, I can soon remedy that!

M: No! No! You give us enough but they give us too much! Chinese teachers give us lots of exams too.

N: Well, I suppose that helps you to pass but I'm more concerned about getting you to communicate.

M: Western teachers are more open-minded too.

N: What do you mean by that?

M: They talk about different topics and have different views about everything and do things in different ways.

N: So tell me which do you prefer?

M: Both, actually. I can learn different things from them.

习惯用语 1

① feel at ease: 感觉自在、不拘束

N: 你既有中国老师又有西方的老师，能跟我说说他们有什么不同吗?

M: 我觉得中国老师更加严厉，没什么幽默感。他们从不会像你那样讲笑话。

N: 可能他们不知道。我喜欢让我的学生感到放松、自在。还有吗?

M: 他们总是给我们留很多作业。

N: 我的作业太少了，得马上改进。

M: 不不! 你的已经足够了，他们留得太多了。中国老师还老是让我

们考试。

N: 我想这也有助于你们通过考试。但我更关注与你们进行交流。

M: 西方的老师思想都很开放。

N: 什么意思?

M: 他们谈论不同的话题,对每件事情都有不同的观点,也有不同的处理方式。

N: 那你更喜欢哪一个?

M: 都喜欢。从二者中,我能学到不同的东西。

Questions 1

1. Who was the best teacher you had? Why?
2. Can you think of any more differences between Chinese and Western teachers?

Dialogue 2

Maggie: What time do your classes start in the morning?

Nick: The first ones would start at nine.

M: So late! You know that ours start at eight.

N: So early! I don't think that British students would ever get up that early!

M: We are used to having classes early and also in the evening.

N: We never have classes in the evening or at weekends.

M: We also have extra classes apart from our major.

N: We don't. So what extra classes do you have?

M: We have mathematics because that is supposed to help us to think more logically.

N: Does it work?

M: Of course! And then there is world history, politics and **PE**①.

N: PE in Britain is voluntary. In fact, every Wednesday afternoon is free so that students can take part in sports if they want to.

M: And then we have optional classes from 7 pm to 9 pm when we can choose from over two hundred subjects.

N: With all those classes it's no wonder that some of my students fall asleep in class!

习惯用语 2

① PE：（Physical Education 的缩写）体育课

M: 在英国，你们早上几点上课？

N: 第一节课 9 点开始。

M: 那么晚！我们 8 点就开始了！

N: 那么早！我觉得英国学生都不可能起那么早！

M: 我们总是早上有课，晚上也是如此。

N: 我们从来不在晚上和周末上课。

M: 除了专业外，我们还有别的课程呢。

N: 我们没有。别的课程是什么？

M: 数学，因为这有助于我们的思维更具逻辑性。

N: 有用吗？

M: 当然了！还有世界史、政治和体育课。

N: 体育课在英国是自愿的。事实上，每周三下午我们都没课，学生们可以参加自己喜欢的运动。

M: 另外，我们晚上 7 点到 9 点还有选修课。从两百多个科目里挑选。

N: 上那么多课，难怪我有些学生会上课睡觉呢！

Questions 2

1. How many classes a week do you have?
2. Which classes would you like to get rid off?

Dialogue 3

Nick: What do you do if you don't like a teacher?

Maggie: We usually give them an evaluation every year so if we don't like someone then we give them a low mark.

N: Wouldn't the teacher be really upset?

M: Sometimes. Of course, he might **get his own back**[①] by giving us low marks too!

N: I know my students complain about some poor teachers and say that the university should replace them with more foreign teachers. They also say that they would like to choose their teachers.

M: I think that would be a good idea. What do you think?

N: I tell them if they do that then it's only fair that I should choose my students too! After all, we want the best teachers to have the best students.

M: I know that some universities have a policy of expelling the worst performing students in order to keep their standards high.

N: Some universities like to make the exams as easy as possible so that as many students as possible pass and that looks good for the university.

M: Many students say that although it's hard to get into university once you're in it's difficult to fail. What about the U. K. ?

N: We have quite a high drop out rate of about 16 percent but that's due to many different factors. Perhaps because it's easy to get in we don't take it as seriously as you do.

M: Yeah, getting to university means that we're more likely to have a good job in the future.

习惯用语 3

① get sb.'s own back: 报复，复仇

N: 要是你不喜欢某位老师的话，你会怎么办？

M: 我们每年都会给老师打分，要是我们不喜欢的话，就给他打低分。

N: 那老师岂不是会很沮丧？

M: 有时候是。当然，他也可以给我们打低分，以示报复。

N: 我知道我的一些学生就在抱怨那些差劲的老师，说学校应该用外籍教师取代他们。他们还说，他们愿意自己选老师。

M: 那倒是个好主意。你觉得呢？

N: 我告诉他们，要是他们那样做的话，我也要挑选我的学生，这样才公平！毕竟，我们想让最好的老师教最好的学生。

M: 我知道有些学校制定了政策，对学生实行末位淘汰制，以保持他们的高水平。

N: 有些学校总是把考题弄得很简单，学生的通过率很高，这样看起来就很好。

M: 很多学生都说，尽管进入大学很难，可一旦你进去了，想不及格也很难。在英国的情况怎么样？

N: 我们的淘汰率很高，大概有 16%。当然，这也取决于不同的因素。也许是因为入学比较简单，我们并不像你们这样重视。

M: 是啊，上大学就意味着将来能找一份好工作。

Questions 3

1. What do you do if you don't like a teacher?
2. Should students be allowed to choose their teachers?

9 食品

‖ Background Information ‖

During a lifetime the average person eats 35 tons of food.

When the famous explorer Marco Polo returned to his homeland of Italy, from China in 1295, he brought back a recipe (among other things). The recipe was a Chinese recipe for a dessert called "Milk Ice". However, Europeans substituted cream for the milk, and so ... "Ice Cream". Ice cream has been a hit ever since! Frankfurter sausages were first created in China.

平均每人在一生当中会吃掉35吨食物。

1295年，当著名探险家马可·波罗从中国返回他的家乡意大利时，他带回去了一个饮品配制方法，这是一道中国甜点的制作方法，当时叫做"奶冰"。不过，后来欧洲人用奶油取代了牛奶，于是就有了"冰激凌"。自从那时起，冰激凌就非常受欢迎。法兰克福香肠最初也是由中国人烹制出来的。

背景信息

Dialogue 1

Nick: Welcome to Fish Nation, Lara.

Lara: Thanks, Nick. I'm so happy that you invited me to this British restaurant.

N: Let's order. Of course, you have no choice. There's only fish 'n'[①] chips!

L: What's so special about it, anyway?

N: We like to think it's our national dish in England. It's normally eaten in the street and traditionally was wrapped in old newspapers.

L: Why old newspapers?

N: That's just wrapping paper that keeps the food hot and enables you to hold it in your hands but now they come in plastic containers.

L: The food's here! What do we do now?

N: Sprinkle plenty of vinegar and salt on them.

L: These chips are so big!

N: Yeah, British chips are larger and thicker than American French fries.

L: I like the fish. Hey, there's no bones!

N: That's why I like it. I really hate bones. You take a mouthful and then you have to chew it carefully so you don't swallow any bones.

L: What's the fish covered in?

N: It's called batter. It protects the fish when it's deep fried.

L: It certainly tastes good. I'll come here again with some of my classmates.

习惯用语 1

① 'n': 〈ṇ〉 = and

N: 欢迎来到 Fish Nation，Lara。

L: 谢谢你，Nick。很高兴你能邀请我来这个英国餐馆。

N: 点菜吧，当然，你别无选择，这里只有鱼和炸薯条。

L: 有什么特别的吗？

N: 我们都喜欢认为这是典型的英国食品。一般都是在街上吃，传统的方法是用旧报纸包着。

L: 为什么是用旧报纸？

N: 就是用它来包裹食物，省得凉了，也便于你用手拿，但现在我们都用塑料盒装了。

L: 食物上来了！我们现在该怎么办？

N: 在食物上撒醋和盐。

L: 这些薯条可真大！

N: 是的，英国的炸薯条比美式炸薯条个头要大。

L: 我喜欢这个鱼，没有刺！

N: 这就是我喜欢的原因。我真的很讨厌鱼刺，你吃一大口鱼，然后得小心翼翼地咀嚼，以免被刺扎到。

L: 这鱼上裹了层什么东西？

N: 这是一层用面粉、鸡蛋、牛奶等调制成的面糊。它可以在炸鱼的过程中对鱼起到保护作用，以免被炸糊。

L: 味道真不错。我要跟我的同学再来一次。

Questions 1

1. Have you ever eaten any British food?
2. Do you like fish with bones? Why? Why not?

Dialogue 2

Lara: What's the main difference between Western food and Chinese food?

Nick: Obviously one difference is the fact that you use chopsticks and we use knives and forks. This affects how the food is prepared and served.

Food and Drink 食品和饮品

L: What do you mean by that?

N: As far as preparation goes because chopsticks can only pick up little bits then Chinese food is cut up into **bite-sized pieces**[①]. We don't have to let the chefs do that because we can use our knives. So our food and especially meat comes in larger portions.

L: I once read that using chopsticks help exercise many joints and muscles which is why we are **manually so dextrous**[②]. What do you mean by served?

N: Well, our food comes on large individual plates whereas you would have several dishes placed in the middle of the table and everyone would help themselves.

L: That's right. We share things in a kind of family way whereas you have your own individual servings. Any other differences?

N: Western food isn't just Western food but global. For example, the most popular food in England is curry which is from India. We have a huge variety of food from all over the world to choose from.

L: In China we have eight regional cuisines as well as food that represents our 56 ethnic groups so you could say that we have an enormous variety too but it's of the domestic kind.

N: I always think that Beijing food is a bit salty.

L: Perhaps it is but that is one of the five fundamental flavours. The others are sweet, sour, bitter and spicy.

N: I know that Sichuan food is spicy. In fact, it's too hot for me!

L: Me too, sometimes! I think you've got to be born in Sichuan to be able to eat their spicy food!

习惯用语 2

① bite-sized pieces: 小块儿的、很容易放进嘴里的食物
② manually so dextrous: 手巧

L: 中餐和西餐的最大不同是什么?

N: 最明显的不同就是你们用筷子,而我们用刀和叉。这就决定了食物的烹调方式和进餐的方式有所不同。

L: 什么意思?

N: 在烹制过程中,由于筷子只能夹起小块的食物,所以中餐的食物都被切成容易入口的小块儿。而西餐厨师就不必这样,因为我们可以用自己的刀来切。所以,西餐的食物,尤其是肉,都是很大块儿的。

L: 我曾经读过一篇文章,说用筷子可以锻炼手上的肌肉和关节,所以我们的手都很巧。你刚才说"进餐的方式"是什么意思?

N: 噢,我们的食物都是放在个人单独的大盘子里面,可你们的菜是放在桌子中间,大家自己夹菜吃。

L: 没错。我们像家里人那样一起用餐,而你们则是自己吃自己的。还有什么不同吗?

N: 西方的食品更具全球性。比如说,在英国最流行的食品是来自印度的咖喱。我们可以选择来自世界各地的各种食品。

L: 在中国,我们有八大菜系,还有来自 56 个民族的食品。所以,我们也有很多选择,不过都是来自一个国家的。

N: 我总是觉得北京的菜有点儿咸。

L: 可能吧,但这是五个基本味道之一。其余的是甜、酸、苦、辣。

N: 我知道四川菜很辣,而且对我来说是太辣了。

L: 有时候我也有这种感觉。我估计只有四川人才能习惯这种辣的食物。

Questions 2

1. Can you use a knife and fork?
2. Which is your favourite Chinese food? Why?

Dialogue 3

Lara: Is it true that British food is really bad?

Nick: I have to admit that in the past it was but over the last twenty years British food has become among the best in the world.

L: Really? I can't believe you!

N: It's true. In a recent survey 15 percent of the world's top restaurants were British!

L: How has that come about?

N: We have many popular cooking shows on TV which have helped but I think it's mainly due to London being the commercial and cultural centre of Europe. We have many international visitors and restaurants have upgraded to cater for them.

L: I've heard a lot about the traditional British breakfast. Can you tell me more?

N: It used to be a great fry-up. Bacon, eggs, sausages, tomatoes and bread were all fried. Very greasy but filling. All **washed down**① with gallons of tea! But now a lot of people have switched over to a more healthy diet such as muesli and toast and orange juice.

L: How are things cooked?

N: Most of our vegetables are boiled and perhaps boiled too much so that they lose a lot of flavour. I must admit I like my vegetables Chinese-style which is stir-fried as they arrive on the plate still crisp.

L: What about meat?

N: We use ovens a lot so most meat is roasted but microwaves are very popular too.

习惯用语 3

① wash down: (用水等) 吞下 (以助消化)

L: 英国菜真的很难吃吗?

N: 我得承认,过去确实是这样,但在过去的 20 年间,英国菜已经变成了世界上最好的菜品之一。

L: 真的? 我可不信!

N: 真的。最近的一项调查显示,世界上15%的顶级餐厅都是英国的。

L: 这个成就是怎么取得的?

N: 我们电视上有很多时尚烹饪节目,这起了不小的作用。不过,我觉得主要还是因为伦敦是欧洲的商业和文化中心。我们有来自世界各地的观光者,所以我们的餐厅需要不断升级,以迎合游客的口味。

L: 我听说过不少关于传统英式早餐的事儿。你能够给我多讲讲吗?

N: 以前主要就是各类油炸食物,比如把培根、鸡蛋、香肠、西红柿还有面包都煎了。油特多,但是很解饿。边吃边喝茶以助消化。但现在人们都倾向于更健康的食品了,比如牛奶什锦早餐、烤面包、橙汁。

L: 食物都怎么烹制?

N: 我们的蔬菜大多是用水煮,可能煮的时间太长而去掉了本来的味儿。我得承认,我还是更喜欢中式蔬菜用旺火炒的方法,菜上来的时候还很清脆。

L: 肉呢?

N: 我们比较常用烤箱,所以大多数肉都是烤出来的,微波炉也很常用。

Questions 3

1. Do you like to watch cookery shows on TV?
2. Can you cook?

10 餐厅

Restaurants

Background Information

London's first Chinese restaurant opened in 1908.
Top Chinese Restaurant Favourites
1. Crispy Duck
2. Sizzling Cantonese Beef
3. Dim Sum
4. Szechuan Prawns
5. Chilli Beef
6. Chicken with Cashew Nuts
7. Lemon Chicken
8. Beef in Oyster Sauce
9. Prawns in Black Bean Sauce
10. Spare Ribs

伦敦的第一家中餐馆开业于 1908 年。
排名前十位的、最受喜爱的中餐菜分别是:
1. 脆皮鸭
2. 广味烤牛肉
3. 点心
4. 白灼虾
5. 水煮牛肉
6. 腰果鸡丁
7. 柠檬鸡
8. 蚝油牛肉
9. 红烧大虾
10. 红烧排骨

背景信息

Dialogue 1

Nick: I've been invited to a Chinese banquet tonight. Can you tell me what will happen?

Iris: If it's in a restaurant then you will probably be in a private room with a large round table.

N: Why is that?

I: It is for privacy so other diners won't interfere and also because it recreates a family atmosphere.

N: We like to be seen when we're eating in public.

I: Any special guests or the host will be seated furthest away from the door. There might be a special napkin arrangement for the special guest too.

N: So that makes them feel special.

I: To show respect for the guests there would normally be eight cold dishes and eight hot dishes.

N: Why the number eight?

I: It's a lucky number in Chinese culture because it's similar in pronunciation to a word that means make a big fortune.

N: We would try to avoid having thirteen guests around a dinner table because that would be an unlucky number.

I: Why?

N: Because there were thirteen people at Christ's last supper, the night before he died.

N: 我今晚受邀参加一个中式宴会，给我讲讲都会发生什么？

I: 如果在餐馆里，那可能会是在一个包间，里面有一张大大的圆桌。

N: 为什么是这样？

I: 为了隐私啊，这样别的食客就不会打扰你们了。同时也是为了营造一种家庭氛围。

N: 在公共场所吃饭的时候，我们不喜欢包间。

I: 特殊来宾或主人会坐在离门最远的座位上。可能还会为特殊来宾

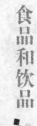

N: 准备特别的餐巾。

N: 这会让他们觉得与众不同。

I: 为了表达对宾客的尊敬，通常会有八道冷菜和八道热菜。

N: 为什么是八道？

I: 在中国文化里，这是个幸运数字。它的发音与汉语"发财"的"发"字相近。

N: 我们会尽力避免在一桌有 13 位宾客，因为这是个不吉利的数字。

I: 为什么？

N: 因为，在基督死去的前一晚，最后的晚餐上就有 13 个人。

Questions 1

1. Do you like to eat in public or private?
2. Are they any other numbers associated with food?

Dialogue 2

Jane: Ling, I feel a little uncomfortable. Why is everyone shouting?

Ling: What do you mean, shouting?

J: Well, it seems like every table is going to fight with each other. I'm especially worried about that guy who keeps hitting his friend so hard on the back.

L: What. I can't see it, what do you mean? Everyone here is so happy, and that table seems to me to be having fun.

J: Fun? But why are they so loud? I thought they are fighting with each other, but you say there is no problem? Why are they being so loud!

L: Oh, calm down. It's nothing about you, they are just happy. Chinese people like to be loud, you know, when we are out we enjoy being loud.

J: OK, I know YOU are loud, but everyone? Why do you need to be so loud? I mean, you are all sitting right next to each other.

L: It's not about space, we just enjoy it. Especially if we are happy, we like to express our feelings aloud.

J: I guess it's the same with the wait staff, when you come in, they all shout "welcome!" and then shout to ask you how many people.

L: What so weird about that? They just want to find a table suitable for you, don't they?

J: In England, when you go inside, a waiter will greet you softly, asking "table for two?" or whatever. We don't like to be so loud.

L: The waiter whispers? How can the people outside see all of the great service? They are trying to sell restaurants, not hide them!

J: Ling, 我觉得有点儿别扭。怎么每个人都在嚷嚷?

L: 嚷嚷? 什么意思?

J: 我觉得每桌的人好像都要打起来似的。尤其是那桌,你看那个人,他一直在用力拍他朋友的后背。

L: 什么? 我没看见啊,你什么意思? 这儿的人都挺高兴的,我觉得那桌的气氛挺高兴的啊。

J: 高兴? 那他们干嘛声音那么大? 我以为他们在吵架呢,你却说很正常。那他们干嘛那么大声说话啊!

L: 噢,别急。这与你无关,他们挺高兴的。知道嘛,中国人就喜欢在外面大声说话,我们喜欢这样,显得热闹。

J: 好吧,我知道你爱大声说话,可是别人呢? 你们有必要那么大声吗? 我的意思是,大家都坐得很近啊。

L: 不是距离的问题。我们就是喜欢这样。尤其当我们高兴的时候,我们喜欢把自己的感觉大声地表达出来。

J: 我估计门口的服务员也是这样,进门的时候,他们都大声说"欢迎光临"。然后大声地问"您几位"?

L: 这有什么好奇怪的? 她们只是想帮你找一张合适的桌子罢了。

J: 在英国，当我们走进饭馆的时候，服务员是很小声地向我们问好，然后会问："你是要两个人的桌子吗?" 或者别的什么。我们不喜欢这么大声说话。

L: 服务员跟你打耳语? 那外面的人怎么能知道里面的顾客接受的服务好不好呢? 他们要推销自己的餐馆，不是隐藏自己。

Questions 2

1. Do you make a lot of noise with your dining companions?
2. Do you shout at the waiters?

Dialogue 3

Tori: Hey, don't look so worried.

Ron: (whispering) I don't know what to do with all these knives and forks and glasses. What do I do?

T: Ah, so that's the problem, don't worry, it's easy. Use the cutlery from the outside to the inside. After each course the dirty dishes and cutlery will be taken away. If you get stuck, just watch someone else.

R: So, I guess the different glasses are for different kinds of drink. OK, OK. I think I got it. Could you explain some more about etiquette? I don't want to do something that looks really stupid!

T: Well, the most important thing is not to make a noise when you eat. Parents always tell their children "don't talk with your mouth full" and "never chew with your mouth open".

R: OK, so that's a bit different. Sometimes it's polite in Asia to make a noise, in my family we slurp our soup and noodles and we are not afraid of burping. What else?

T: At a formal meal, in a place like this, you should always have your napkin on your lap and don't rest your arms on the table either. If you are at someone's house, it is always polite to compliment the cook. I guess that last part is the same everywhere in the world.

R: Well, although you didn't cook it, thanks for a great meal!

T: It's a good job I didn't cook it, I can burn water!

R: Oh, what do I do if I need to leave the table? Do I need to say where I'm going?

T: Don't go into too much detail. The safest option is to just say "please excuse me for a minute".

R: OK, well, in that case. Please excuse me for a minute.

T: 嘿, 你怎么看起来一副焦急的样子。

R: (小声说) 我看到这么多刀、叉还有杯子就不知所措。这些东西怎么用?

T: 噢, 原来是因为这个, 别急, 很简单的。按照由外到内的顺序使用餐具。每吃完一道菜, 用过的餐具就会被撤走。要是不明白的话, 就看看其他人是怎么用的。

R: 那这些杯子是为了喝不同的饮料准备的吧? 嗯, 我想我明白了。再给我讲讲用餐礼节方面的注意事项好吗? 我可不想冒傻气。

T: 嗯, 最主要的就是不要在吃东西的时候发出声音。家长们也会经常对孩子说 "吃东西的时候不要说话", "咀嚼东西的时候不要张嘴"。

R: 好吧, 看来还真有些不同。在亚洲, 有时候吃饭发出声音表示一种礼貌。在我们家, 喝汤、吃面条的时候都会发出啧啧声, 打嗝也可以。还有别的吗?

T: 在正式场合用餐的时候, 比如在这种地方, 要始终保持餐巾在你的膝盖上, 而且也不要把手放在桌子上休息。要是在别人家里用餐, 要不时地对饭菜进行称赞, 以示礼貌。我想后面这点在世界各地都是一样的。

R: 好的, 尽管这菜不是你做的, 但这顿饭真的很不错。

T: 我不会做饭, 但我可以烧开水啊。

R: 那要是我想离开一下怎么办? 我有必要告诉他们我要去哪里吗?

T: 不用说的那么具体。最保险的办法就是说"不好意思，我稍微离开一下"。

R: 噢，那么，不好意思，我要稍微离开一下。

Questions 3

1. Do you know how to use cutlery?
2. Do you slurp your food?

11 小吃

Snacks

Background Information

Snacks are food that is meant to relieve your hunger but not take it away completely. They are often eaten in between meals. In the U. K. the number one snack is the sandwich and we eat 11. 5 billion every year which means about 200 each.

In Beijing there is the Donghuamen Night Market which is just next to Wangfujing. Here you can find just about anything on a stick from insects and worms to seahorses.

小吃就是那种用来临时解饿的东西，吃不饱。通常都是在两顿正餐的中间吃。在英国，最流行的小吃是三明治，我们每年会吃掉115亿个，也就是每人每年要吃掉200个。

在北京，有个东华门夜市就在王府井附近，那里卖的小吃从小昆虫到海马，卖什么的都有。

背景信息

Dialogue 1

Derek: I'm so glad you convinced me to come out tonight. I needed the break, work has been so busy recently.

Mary: Yes, I know so I thought this would be the perfect place for you to relax. We can do your two favourite things, walking and eating.

D: Night markets are great. We don't have anything like this back home, I mean, we have street vendors selling hot dogs and burgers and stuff, but the variety here is massive.

M: Also, everything is easy to eat while you walk. It's all on a stick so you can get a snack but still go shopping or whatever.

D: Yeah, it's a great idea. Everything on a stick! Certainly handy! What's this one? It looks different, is it a kind of meat?

M: No, it's called Yunnan popcorn. It's really yummy. I think you should buy one, just try it. You'll be surprised.

D: OK, I'll give it a go.

M: So, what do you think? Tasty, right?

D: Yes. But what is it?

M: Bamboo worms.

D: Worms! I'm shocked! It looks so gross, but it tastes gorgeous. Thanks for recommending it to me. I still can't believe it's an insect, though.

M: Would you like to try these? They're scorpions or you could try the silkworms or the seahorses?

D: I'll try anything once! But it just seems weird eating insects.

M: It's not too weird. You know, I'm from Guangdong. We have a saying about Cantonese people, we will eat anything with legs, except a table and anything that flies, except a plane and anything that swims, except a submarine!

D: 我真的要感谢你今天晚上把我叫出来。我需要休息了，最近工作实在是太忙了。

M: 我知道，所以我觉得这个地方应该很适合让你放松一下。我们可以散步、吃东西，这都是你喜欢的。

D: 我特喜欢逛夜市。在我们那儿，可没有这样的夜市。我们只有街边摊贩，卖些热狗、汉堡之类的东西。而你们这儿吃的可是太丰富了。

M: 而且这些东西还特别适合边走边吃。小吃都是插在签子上的，可以一边吃，一边买东西或者干别的什么。

D: 嗯，是个好主意。所有的东西都插在签子上，真方便。这个是什么？看起来不太一样，是肉吗？

M: 不是，这个叫云南爆米花，特别好吃。我觉得你应该买一个尝一尝。肯定会大吃一惊的。

D: 好吧，我尝尝。

M: 怎么样？味道不错吧？

D: 嗯，但这个是什么？

M: 竹叶虫。

D: 虫子！天哪！看上去很恶心，但吃起来味道真不错。谢谢你给我推荐这个好吃的。但我还是不相信这是一种虫子。

M: 想尝尝这个吗？蝎子、蚕蛹，或者尝尝海马？

D: 我都会尝一下！但我觉得吃昆虫挺奇怪的。

M: 没什么可奇怪的。我来自广东，对我们广东人有一种说法，除了桌子，有腿的都吃；除了飞机，能飞的都吃；除了潜艇，能游的都吃。

Questions 1

1. What kind of insects do you like to eat?
2. What kind of insects do you not like to eat?

Dialogue 2

Fred: What's that you're eating? It looks very interesting.

Rita: It's called "tanghulu". It's a famous local snack. There are

Food and Drink 食品和饮品

many different kinds, this one is crab apple.

F: Yes, it looks like lots of mini toffee apples. Have you heard of toffee apples?

R: No, but I'm guessing they are made the same way. With lots of sugar.

F: Yes. But toffee apples are just apples, we don't usually have any other kinds of fruit on a stick.

R: Well here there are many different kinds to choose from. Look, there's another vendor over there.

F: Wow! What a selection, I can see the crab apple ones, but also haw, pineapple the list is endless.

R: I think you should try the haw one, it's really sweet and I know you have **a sweet tooth**①.

F: OK, if it's sweet I'm bound to like it. I've noticed here that everything seems to be on a stick!

R: In China we put lots of things on sticks. This means we can eat it without having to touch it.

F: Oh, I see. It's a hygiene thing. It's a great idea. In Western countries, we usually eat things like sandwiches and crisps **on the go**②. I never considered it as being unhygienic, but I guess it is. We have to touch the food while we eat it.

R: So, if you put it on a stick, you don't need to touch it. After you have finished, you simply throw the stick away.

习惯用语2

① a sweet tooth: 喜好吃甜食　　② on the go: 走路的时候

F: 你吃什么呢？看起来还挺有意思的。

R: 这个东西叫"糖葫芦"，是一种很有名的地方小吃。它的种类很多，这个是海棠的。

F: 是啊，看起来就像很多小的"太妃苹果"。你听说过"太妃苹果"吗？

R: 没有，但我想跟这个的做法也差不多。都用很多糖。

F: 没错，但"太妃苹果"就只是苹果。我们一般不会用别的水果作原料。

R: 哦，糖葫芦的原料可就多了，看，那边又有一个卖糖葫芦的。

F: 哇！种类这么多！除了海棠的还有山楂的、菠萝的……好多种啊！

R: 我想你应该尝尝山楂的。那个很甜，我知道你爱吃甜食。

F: 好吧，要是很甜的话，我肯定喜欢。我发现这里的小吃似乎都是串在签子上的。

R: 在中国，我们把很多小吃都串在签子上。这样我们不用手碰也能吃到了。

F: 哦，我明白了。这倒挺卫生的，是个好主意。在西方国家，我们经常边走边吃东西，比如三明治、炸薯片啊什么的。我从没觉得这样会不卫生，不过，可能还真有点儿。因为我们吃的时候得用手碰吃的东西。

R: 所以，要是把它插在一根签子上，就不必用手碰了。吃完只要把签子一扔就可以了。

Questions 2

1. What's your favourite snack?
2. How often do you eat snacks in a week?

Dialogue 3

Elaine: What kinds of snacks do Westerners eat?

Nick: We like things like crisps, biscuits and of course chocolate!

E: They all sound very fattening!

N: That's the problem. They are! We like to eat snacks while we're watching TV or a movie.

E: What kind of snacks do you eat in the evening?

N: Donner kebabs are very popular. This is a kind of Turkish food and consists of slices of pork with salad in Turkish bread. It's common to eat one after a night of drinking.

E: A kind of sandwich then?

N: We are really fond of sandwiches. You know, it was invented by the Earl of Sandwich who apparently didn't want to leave the **gaming table**① so he ordered two slices of bread and put some meat between them and that is how the sandwich came into being.

E: We like sandwiches too. You know that Subway shops are popular here.

N: Yeah. My favourite is a **BLT**②.

E: I like a **Club**③ because there's practically everything you want in it!

N: You're making me feel hungry. Let's go to Subway now, shall we?

E: Good idea!

习惯用语 3

① gaming table: 赌桌
② BLT: 熏肉、生菜、西红柿三明治
③ Club: 一种三明治，里面含有鸡胸脯肉和培根，拌有西红柿、生菜，还配有蛋黄酱。

E: 西方人喜欢吃哪些小零食？

N: 我们喜欢薯片、饼干，当然还有巧克力。

E: 这可都是容易发胖的东西啊。

N: 没错！这就是问题所在。我们还喜欢边看电视或电影边吃零食。

E: 你们每天晚上都吃什么？

N: 多诺肉串很流行，它是一种土耳其食品，用土耳其面包夹着沙拉拌猪肉片。我们常常在喝了一夜酒之后吃一块儿。

E: 是一种三明治吗？

N: 我们真的很喜欢吃三明治。你知道嘛,这个东西是三明治伯爵发明的,他很爱赌钱,不想离开赌桌,所以就点了两个面包片然后把肉夹在中间,三明治就这样被发明出来了。

E: 我们也喜欢三明治。赛百味在这里很流行。

N: 哦,我最喜欢的是熏肉、生菜、西红柿三明治。

E: 我喜欢克拉布三明治,因为它里面几乎什么都包括了。

N: 你一说,我都觉得饿了。咱们去赛百味吧,怎么样?

E: 好主意!

Questions 3

1. Do you like to eat sandwiches?
2. What's your favourite sandwich filling?

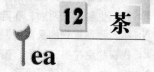

12 茶

Background Information

The drinking of tea is popular in both China and the U. K.

Tea is often thought of as being a quintessentially British drink, and we have been drinking it for over 350 years. But in fact the history of tea goes much further back.

The story of tea begins in China. According to legend, in 2737 B. C., the Chinese emperor Shen Nung was sitting beneath a tree while his servant boiled drinking water, when some leaves from the tree blew into the water. Shen Nung, a renowned herbalist, decided to try the infusion that his servant had accidentally created. The tree was a Camellia sinensis, and the resulting drink was what we now call tea.

不论是在英国还是中国，喝茶都很流行。

茶，一直被认为是英国饮品中的精粹，至今我们已经有350多年的饮茶历史了。但事实上，茶的历史要更加久远。

茶的历史起源于中国。据传说，公元前2737年，当时的帝王神农正坐在一棵树下休息，他的仆人在给他煮开水喝。这时候，有几片树叶吹落进了水中。作为当时最著名的中医，神农决定尝一尝他的仆人偶然间创造出的饮料。这棵树就是茶树，而他饮用的就是我们现在所说的茶。

背景信息

Dialogue 1

Hilary: Do British people drink a lot of tea?

Nick: Tea drinking is part of our culture. During work we would have an official tea break in the morning and afternoon although now it is mostly coffee that is drunk. It used to be our national drink but now we drink more coffee than tea.

H: Is there any special English tea ceremony like we have with Chinese tea?

N: Good heavens[①], no. We just pour a little hot water into a teapot first to warm it, empty it, put some tea leaves in, add hot water and then leave it for a few minutes to brew.

H: Is English tea different than Chinese tea?

N: It actually comes from India and we regard it as black tea but you would probably call it red tea!

H: I have sometimes had some English tea called Lyons but it was very strange as it came in a bag!

N: Tea bags are very popular in Britain as we don't like the leaves floating about in the tea. Did you add milk and sugar to the tea?

H: No, of course not! What a strange thing to do!

N: But that is exactly how we like our tea! We think it improves the taste.

H: Not for me, it doesn't! Do you drink a lot of English tea in China?

N: No, I have got used to Chinese tea so now I drink plenty of green tea.

习惯用语 1

① Good heavens: （表示惊讶、吃惊）天哪

H: 英国人经常喝茶吗？

N: 饮茶是我们文化的一部分。工作的时候，我们通常都会在上午和下午分别有一次饮茶休息时间，但现在大多都被喝咖啡取代了。原来，茶在全国范围内都很受欢迎，但现在我们更多的是喝咖啡。

H: 英国人喝茶是否像我们中国人喝茶一样，有什么特殊的讲究吗？

N: 天哪，没有。我们只是往茶壶里灌一些热水暖一下，然后把水倒掉，加入茶叶，放入开水，盖上盖子焖几分钟，把茶叶完全泡开。

H: 与中国茶比起来，英国茶有什么不同吗？

N: 英国的茶来自印度，我们称其为黑茶，你们好像更愿意称其为红茶。

H: 我喝过一种英国茶叫里昂。但很奇怪，它是放在袋子里的。

N: 袋装茶在英国很流行，因为我们不喜欢茶叶飘浮在杯子里。你们会往茶水里面添加牛奶或糖吗？

H: 当然不了！这太奇怪了！

N: 但我们就喜欢这样喝茶！这样喝起来味道更好。

H: 我可不这么觉得。你在中国也经常喝英国茶吗？

N: 不，我已经习惯中国的茶了，现在我喝得更多的是绿茶。

Questions 1

1. Do you drink a lot of tea?
2. Have you ever tried British tea? If so, what did you think of it?

Dialogue 2

Nick: I've noticed that Chinese people often drink hot tea even on scorching sunny days when we would have cold drinks. Why is that?

Hilary: We believe that heating the inside of the body balances the temperature and helps to mediate the external heat. It not only quenches the thirst but also makes you feel cool. In the same way we would eat ice cream in winter.

N: I suppose that makes sense but it still seems odd to me.

H: Look at it another way. If a person faints then you might douse them with cold water which shocks the body's system and helps to revive that person. We try not to shock the body's system, that's all.

N: Sounds logical when you put it that way but we're still used to having ice-cold drinks in the summer and hot drinks during winter. But isn't Chinese tea also good for the health?

H: We think so, yes. You know that China is the home of tea and it has been popular ever since the fourth century. We have an old saying that "tea tastes bitter and to drink it improves the mind, dispels laziness, enlivens the body and brightens the sight."

N: Sounds good to me. Maybe I should give up coffee!

H: Drinking tea, especially Oolong tea, is also good for losing weight.

N: I should definitely start drinking tea then!

H: It's also good for minor ailments and helping digestion.

N: You've convinced me! From now on, it's tea, tea, tea!

N: 我发现中国人即使在炎热的夏季，该喝冷饮的时候也喝热茶，这是为什么啊？

H: 我们认为喝热茶水可以加热身体内部，平衡体温，有助于调节适应外面的热度。不但可以解渴，还可以让我们觉得凉爽。同样的，我们也会在冬天吃冰激凌。

N: 我想这也许有点道理，但我还是觉得这很奇怪。

H: 咱们换个角度来看。要是一个人晕倒了，你们就可能用冷水泼他，来刺激他的身体系统，以帮助他苏醒。而我们只是为了不去刺激身体系统罢了。

N: 要是这么说的话，还有点儿道理，但我们还是习惯夏天喝冷饮，冬天喝热饮。但是，中国茶不是对人体也有好处吗？

H: 是的。中国是茶的故乡，从公元4世纪就已经流行了。我们有一句老话："浓茶可以清醒神志，驱散懒散，使身体充满活力，还能明目。"

N: 这听起来不错。也许我该放弃咖啡，改喝茶了。

H: 喝茶，尤其是乌龙茶，还有利于减肥。

N: 那我现在就该开始喝茶！

H: 喝茶还可以治疗小的疾病，帮助消化。

N: 你真是把我说动了！从现在起，我只要茶，茶，茶！

Questions 2

1. Do you drink Chinese tea because of its health qualities?
2. How much Chinese tea would you drink every day?

Dialogue 3

Hilary: Is Chinese tea popular in Britain?

Nick: Very popular. I think people like it because it's viewed as being a healthy drink. It's relaxing, has low levels of caffeine and plenty of **antioxidants**①.

H: You know we have many kinds of Chinese tea. What kinds are popular in the U.K.?

N: I think green tea. I don't think that British people know too much about it. For example, they would not know that the best green tea comes from Hangzhou and is called Longjing tea.

H: Is it cheap to buy?

N: Most kinds are but I heard that one tea shop in Edinburgh sells a pot of silver needle white tea for twenty pounds.

H: That is a lot of money but I do know why it is expensive. It is picked for just two days a year and it is just the top two leaves that come off. It's called needle tea because it's a long, thin rolled leaf.

N: At Harrods, the famous department store in London, there is a green tea that costs about nine pounds a cup so obviously some people are becoming sophisticated tea connoisseurs.

H: We actually have seven kinds of tea. There's green tea of course, jasmine tea, Oolong tea, keepfit tea, black tea, white tea and brick tea.

N: What's brick tea?

H: It's made from tea dust or inferior tea and pressed into blocks, usually round shaped. The most famous one is Pu'er tea which comes from Yunnan.

N: I must try that one sometime!

习惯用语 3

① antioxidant: 抗氧化剂（某些维他命，比如维他命 E，就是抗氧化剂，可以保护人体防止因自由基氧化而产生的影响）

H: 中国茶在英国很流行吗？

N: 很流行。我觉得人们之所以喜欢它，是因为它被视作一种健康饮品。它可以让你放松精神，咖啡因含量极低，但抗氧化剂含量很高。

H: 你知道中国茶的品种很多。在英国，哪种最流行？

N: 我觉得是绿茶。英国人似乎对喝茶并没有什么研究。比如，他们并不知道最好的绿茶是来自杭州的龙井茶。

H: 价钱便宜吗？

N: 大多数的价格还算便宜，不过我听说爱丁堡有个茶庄，卖一罐银针白茶要 20 英磅。

H: 那可够贵的，但我知道为什么会这么贵。这种茶叶每年只有两天的采摘期，而且只要最顶端的两片叶子。之所以叫它针茶，是因为茶叶的形状细而长。

N: 在伦敦最著名的百货商场 Harrods，有一种绿茶要 9 英镑一杯，所以，显然人们对喝茶可谓是越来越讲究了，都成为饮茶行家了。

H: 其实，中国茶叶分为七大类，有绿茶、茉莉花茶、乌龙茶、保健茶、红茶、白茶和砖茶。

N: 什么是砖茶？

H: 就是把茶叶末或者次等茶压缩而成，通常是圆形的。最著名的是来自云南的普洱茶。

N: 有机会我得尝一尝。

Questions 3

1. Which kind of Chinese tea do you like best? Why?
2. Have you tried all seven types?

13　文学

Literature

‖ **Background Information** ‖

The four Chinese literature classics are

1. *A Dream of Red Mansions*
2. *The Three Kingdoms*
3. *Journey to the West*
4. *Outlaw of the Marshes*, or *Outlaw of the Water Margins*

There are many British literature classics but perhaps the two most famous are:

1. *The King James Bible* (1611) by William Tyndale and Fifty-four Scholars Appointed by the King
2. *The First Folio* (1623) by William Shakespeare

中国的四大名著是:

1. 《红楼梦》
2. 《三国演义》
3. 《西游记》
4. 《水浒传》

英国也有很多文学名著, 其中最为著名的两部可能是:

1. 《詹姆士国王圣经》 (1611, 威廉·廷代尔, 以及由国王钦点的其他54位专家)
2. 《第一对开本》 (1623, 威廉·莎士比亚)

背景信息

Dialogue 1

Nick: Hi Sally. What are you reading?

Sally: It's about Confucius and his rules for life.

N: Rules for life, like in a religion?

S: Exactly! Confucianism is one of "The Three Ways", along with Buddhism and Taoism.

N: Oh, I see. When did he live?

S: He lived from 551–479 B. C. His life defines the end of The Spring and Autumn period in Chinese history. It wasn't until many years after he died that he became the dominant Chinese philosopher of morality and politics.

N: So, when did he become famous?

S: In the Warring States Period from 390–305 B. C., they extended and systematised his teachings, but they didn't become **what we know today**① until the Han dynasty.

N: So, it was years and years after his death? I guess it's a little like Jesus, whose teachings became the official religion of the Roman Empire 300 years later.

S: But Confucianism is all based on a terminology of morality.

N: What do you mean by that?

S: OK for example, he believed in benevolence, charity, humanity and love. Or overall, kindness, which is called "Ren".

N: To be kind to each other, **that makes sense**②.

S: Yes, and right conduct, morality and duty to one's neighbour. It's called righteousness, or "Yi".

N: Yes, the *Bible* says we should love our neighbours, so that's a little similar.

S: Most religions **have things in common**③. Also, profit, gain and advantage, which are not proper motives for actions affecting others. It's called "Li", and it's why Confucius thought that commerce and industry were bad ideas.

N: I see. So are these the 3 main things?

S: Yes. "Ren" is in the Beings' category, "Yi" the Doing or Means' category and "Li" is part of the Ends' category.

N: Sounds complicated!④

S: A simple way to remember is this to love others; to honour one's parents; to do what is right instead of what is of advantage; to practice "reciprocity," or "don't do to others what you would not want yourself"; to rule by moral example, called "De" instead of by force and violence.

N: OK, so be a nice, kind person, love each other and don't fight.

S: Yes, that's right.

习惯用语 1

① what we know today: 我们如今所了解的
② that makes sense: 听起来有道理
③ have things in common: 在某些方面有共同点
④ Sounds complicated! 听起来很复杂!

N: Sally, 你在看什么书呢?

S: 关于孔子及其人生准则方面的书。

N: 人生准则? 就像宗教信仰一样?

S: 没错! 儒教是"三教"之一, 另外还有佛教和道教。

N: 哦, 我明白了。孔子生活在什么时代?

S: 他生活在公元前551年至前479年, 正是中国历史上的春秋晚期。在他死后很多年, 才被认为是中国最有影响的思想家和政治家。

N: 他什么时候开始出名的?

S: 在战国时期, 就是公元前390年至前305年, 人们将孔子的教义学说充实丰富并加以系统整理。不过, 我们今天所了解到的孔子儒家思想是到汉朝才成形的。

Culture 文化

N: 那这都是他死后很久的事情了。我觉得这有点儿像耶稣，耶稣的教义直到他去世 300 年以后才成为罗马帝国的国教。

S: 不过孔子的儒家学说是以一整套的道德规范为根本基础的。

N: 这么说是什么意思？

S: 比如说，他信奉善心、慈爱、人性与真爱。总之是善良，就是我们所说的"仁"。

N: 与人为善，这说得有道理。

S: 没错，还有对自己身边的人要品行端、讲道德和负责任，这就是道义，或者叫做"义"。

N: 没错，《圣经》上说，我们应该爱自己身边的人，在这点上倒是有点儿相似。

S: 大多数宗教都有一些共同点。此外，利益、收益和好处不能成为你妨碍他人的正当动机，这叫"礼"。所以，孔子认为，从事工商业不好，原因就在于此。

N: 明白了。这就是他的三个主要观点吗？

S: 没错。"仁"，属于"本体"范畴，是对事物的态度；"义"，属于"行为或方式"范畴，是做事的准则；"礼"，属于"目标"范畴，是行为的目的。

N: 这听起来可有点儿复杂。

S: 也可以简单一点儿记，这就是：爱他人；敬父母；做正确的事情而不是唯利是图；要"互惠双赢"，即"己所不欲，勿施于人"；要以"德"服人，而不是靠武力与暴力压人。

N: 嗯，就是要成为一个善良的人，互相关爱，放弃斗争。

S: 是的，没错！

Questions 1

1. Have you ever read Confucius?
2. Did you ever watch the TV series by Yu Dan?

Dialogue 2

Sally: I've been reading Shakespeare's *Romeo and Juliet* but I find it really **tough going**① as the language is so difficult to understand.

Nick: That's true for native speakers too. I know that when I read one of his plays I am always glancing at the glossary and notes so that I can understand the words.

S: So how did Shakespeare become so popular if people can't understand what he wrote?

N: People in his time understand his plays because they were meant to be heard not read. It wasn't until seven years after his death that his plays were published in printed form.

S: But I read somewhere that Shakespeare invented about 2,000 new words. How did they understand these new words if they'd never heard them before?

N: Well some nouns he used as verbs, as adverbs, as substantives, and as adjectives. He also created new compound words such as " baby-faced" and " smooth-faced". He also brought into common use words that had appeared in the last half of the 16th century.

S: Perhaps it is the phrases he is most remembered for such as "To be or not to be, that is the question" and " Romeo, Romeo, wherefore art thou Romeo?" I know that some people don't think that Shakespeare wrote the plays. Why is that?

N: Primarily because he never went to university so some critics don't think that he received a high enough education to be able to write such great literature. Personally, I agree with the majority of Shakespearean academics who think Shakespeare was the author.

S: Well "a rose by another name would still smell as sweet"!

N: That's right. **One thing is for sure**[②] and that is that the world's greatest plays have enriched the English language ever since and his plays are among the finest literature the world has ever seen.

① tough going: 很难读下去
② one thing is for sure: 确定的、有把握的事情

S: 我最近一直在读莎士比亚写的《罗密欧与朱丽叶》，我发现里面的语言太晦涩难懂了，很难读下去。

N: 没错，即使对英国人来说，也很困难。我记得当年我读他的一部戏剧的时候，得不停地翻阅词汇表，看注释，这样才能明白是什么意思。

S: 既然人们都看不懂他写的是什么，那莎士比亚为什么还会这么出名呢？

N: 他那个时期的人能够理解他的作品啊，因为这些作品是让人们听的而不是读。直到他去世 7 年以后，其作品才以书籍的方式出版。

S: 我曾看到过一篇文章说莎士比亚发明了大约 2000 个新词。观众从来没听过这些词，怎么能理解是什么意思呢？

N: 他主要是把一些名词用作动词、副词、独立成分和形容词等等。他还发明了一些复合词，比如"娃娃脸"、"没有胡须的脸"。很多 16 世纪后期才出现的词汇他也经常使用。

S: 或许人们能记住莎士比亚主要是因为这些话，比如"生存还是毁灭，这是个问题"，还有"罗密欧啊，罗密欧，为什么你偏偏是罗密欧？"我知道有些人不相信是莎士比亚写了这些剧本，这是为什么呢？

N: 可能是因为他从没有上过大学，因此一些评论家认为，既然他没有接受过足够好的高等教育，所以不可能写出如此的鸿篇巨著。不过我个人还是倾向于大多数学者的观点，相信莎士比亚就是真正的作者。

S: 嗯，这就叫"玫瑰取别名，芳香依旧存"，这些伟大的剧作是谁写的，这不重要了。

N: 没错。有一点可以肯定的，莎士比亚的剧作堪称世上最伟大的，它们丰富了英语这门语言，而他的这些戏剧无疑也是世界文学宝库中的一朵奇葩。

Questions 2

1. Which is your favourite Shakespeare play? Why?
2. How do you deal with the difficult language?

Dialogue 3

Sally: I've been watching a great old TV show, it's called *A Dream of Red Mansions*, have you heard of it?

Nick: *A Dream of Red Mansions*, no, I don't think so. Why is it so good?

S: Well, it's so good that they turned the filming area into a park in Beijing. We should go there, then you will understand.

N: I'd really like to go. We can explore, but I need to know about it first. Why is an old TV show so interesting?

S: Well, actually it was a book written in the mid 18th century, during the Qing dynasty. It has been popular since then, not just now.

N: So, what's it about? Do you think a foreigner could enjoy it? Or is it **too steeped in history**[①]?

S: *A Dream of Red Mansions* is known to every household in China. We all love it, and nowadays you can get it with English subtitles. So you can understand. It talks about women suffering under a male oriented society.

N: Oh, I see, you should like it then! But what is the main storyline?

S: It is said to be a story engraved on a stone. Legend has it that one day two immortals passed by a huge wild mountain, bringing to the world a stone, which is said to be one of those left by Nu Wa, a girl who used to patch up the sky with stones.

N: So, it's all about stones? That seems like a strange storyline.

S: Not all about stones, not at all. The main idea of the book is

Culture 文化

to see it through the eyes of a couple of young lovers. You should be interested in that, since you have so many girlfriends.

① too steeped in histoy: 太多的历史事实

S: 我最近一直在看一部特别棒的老电视剧,叫《红楼梦》,你听说过吗?

N: 《红楼梦》? 没有,没听说过。它好在哪里?

S: 特别好,所以人们为了拍摄这部电视剧,还专门在北京建了一个公园。咱们到了那儿,你就明白了。

N: 我很想去那儿,可以到处看看,不过首先我得知道它讲的是什么。为什么一部老电视剧如此有趣?

S: 这部电视剧改编自同名小说,是在 18 世纪中期写成的,当时是中国的清朝时期。从那时候起就开始流行了,而不是现在。

N: 讲的是什么? 你觉得一个老外能理解吗? 会不会因为涉及到太多的历史方面而难懂?

S: 《红楼梦》在中国是家喻户晓。我们都很喜欢它,现在你也可以买到英文版本的。这样你就能明白了。讲的是女性在男尊女卑的社会中所承受的痛苦。

N: 哦,我明白了。那你是应该喜欢。故事情节是什么?

S: 据说这个故事原本是刻在一个石头上的。相传有两位仙人路过一座大山,为世界带来一块石头,这块石头就是女娲补天时留下来的。女娲是中国神话中的一个女子,相传是她用石头补天的。

N: 那这个故事讲的就是关于石头啊? 这种故事情节太奇怪了。

S: 不是,根本不是讲石头的。整个故事主要是从一对年轻恋人的视角来观察的。你有那么多女朋友,按理说你应该对此感兴趣。

Questions 3

1. Have you ever read *A Dream of Red Mansions*?
2. Have you seen the film, the TV show and Peking opera of the same name?

14 音乐

Music

Background Information

Chinese music is the body of vocal and instrumental music composed and played by Chinese people. For several thousands of years Chinese Culture was dominated by the teachings of the philosopher Confucius, who conceived of music in the highest sense as a means of calming the passions and of dispelling unrest and lust, rather than as a form of amusement. The ancient Chinese belief is that music is meant not to amuse but to purify one's thoughts. Traditionally the Chinese have believed that sound influences the harmony of the universe. Until quite recently the Chinese theoretically opposed music performed solely for entertainment, accordingly, musical entertainers were relegated to an extremely low social status. Melody and tone are prominent expressive features of Chinese music, and great emphasis is given to the proper articulation and inflection of each musical tone.

中国音乐就是由中国人创作并演奏的、歌唱与乐器相结合的音乐形式。在过去几千年的时间里，中国文化一直都由儒家思想占据支配地位，这种学说认为，音乐就其终极意义来讲，是用来平抑激情、排解烦躁与欲望的方式，而不是将其视为一种娱乐消遣。古代中国人认为音乐不是用来愉悦他人，而是用来净化思想的。一直以来，中国人认为音乐能够影响宇宙万物的和谐。由于中国人从理论上反对将音乐单纯地视为一种娱乐，因此中国音乐表演者的社会地位也异常低下，直到近些年，这种状况才得以改观。中国音乐的主要特点就是旋律和音调，非常强调每个音调的清晰度和音调变化。

背景信息

Dialogue 1

Kitty: Who's your favourite band?

Nick: I would have to say The Beatles. Have you heard of them?

K: Who hasn't? They were the original boy band weren't they?

N: That's right. Four young men from Liverpool **caused a sensation**[①] when they played in their local club called the Cavern. They created Beatlemania where girl fans went wild and would shout and scream.

K: What was so special about them?

N: They wore suits and had long hair and Paul, their lead singer was so handsome. But mainly it was their music.

K: So which of their albums would you suggest I listen to first?

N: You could try their latest album which was released in December 2006. It's called *Love* and is a reworking of some of their **best hits**[②].

K: Love seems to be what their music is all about. Did you know that every song from their first 5 albums was about love?

N: I know their first U.K. single was *Love Me Do* released in October 1962 so that does not surprise me.

K: Which of their albums do you like best?

N: I would have to say *Rubber Soul*. I could **listen to that all day**[③]. The song writing partnership of Lennon and McCartney was the most successful in the world.

K: Yet they never wrote their songs together. One way to tell who wrote a song is to listen to who the main singer is.

N: They were the most successful band in the world from 1963 until 1970 when Paul McCartney left the band to pursue his own musical career.

K: It was really tragic the way John Lennon died, wasn't it?

N: Yes. He was signing autographs when a fan shot him 5 times in 1980. There's a memorial to him outside his home in New York.

K: And then George Harrison died of cancer and now there are only two Beatles left.

N: Ringo Starr and Paul McCartney. Paul's wife Linda died of cancer too. He remarried but unfortunately that marriage ended after **lurid tales**④ of his wife's behaviour in her younger days emerged in the press.

K: I hope he finds true love at last. After all, their music has given us a lot of love, right?

N: Yeah, she loves me yeah, yeah, yeah!

习惯用语 1

① cause a sensation: 引起轰动
② best hits: 销售排行榜首位的唱片
③ listen to that all day: 花一整天的时间来听音乐
④ lurid tales: 可怕的流言蜚语

K: 你最喜欢哪个乐队?

N: 我想应该是甲壳虫乐队了。你听说过吗?

K: 这谁没听说过啊? 他们是早期的男孩组合, 对吧?

N: 没错。来自利物浦的 4 个年轻人在当地一个叫做"深洞"的俱乐部表演, 由此掀起了一股热潮。他们引起了甲壳虫热, 女粉丝们疯狂尖叫并呐喊。

K: 他们有什么特别的?

N: 他们穿西装, 留长发。他们的主唱 Paul 特别帅! 但主要还是他们的音乐吸引人。

K: 那你建议我先听听他们的哪张专辑?

N: 你可以听听他们在 2006 年 12 月发行的最新专辑《爱》, 该专辑收录了当年他们的多首经典曲目, 现在重新演绎了。

K: 他们的音乐主题似乎都是关于爱的。你知道嘛, 他们最早的 5 张专辑中每首歌都是关于爱的。

Culture 文化

N: 我知道，他们第一个在英国发行的单曲叫《爱我吧》，是在1962年10月发行的，所以他们的歌曲以爱为主题就不足为怪了。

K: 你最喜欢他们的哪张专辑?

N: 我想应该是《橡胶灵魂》。我能听上一整天都不感到腻。Lennon 和 McCartney 是世界上最成功的歌曲创作组合。

K: 可是他们从没有一起创作过歌曲。有一个方法可以分辨歌曲是由谁创作的，就是听听主唱是谁。

N: 从1963年开始，他们一直是最成功的乐队。直到1970年，Paul McCartney 离开了乐队，去追寻自己的音乐事业。

K: John Lennon 死得非常惨，对吗?

N: 没错。1980年，他正在签名的时候，一个粉丝向他连开5枪。在他纽约的家门外还有一个纪念碑。

K: 后来 George Harrison 因癌症去世。现在只剩下两名甲壳虫成员了。

N: Ringo Starr 和 Paul McCartney。Paul 的妻子 Linda 也死于癌症。后来他再婚了，不幸的是由于媒体大肆报道他妻子年轻时有不检点的行为，这段婚姻终因谣言而告终。

K: 但愿他最终找到了真爱。毕竟他们的音乐给了我们很多爱，对吧?

N: 是的，她爱我! 耶，耶，耶! (she loves me yeah, yeah, yeah! 是该乐队一首歌的歌词，这里有语带双关的味道。——译者注)

Questions 1

1. Have you ever listened to any Beatles songs?
2. Who's your favourite U. K. band? Why?

Dialogue 2

Dylan: What do you think of My Hero?

George: It's just a **pin-up show**① for girls to look at talentless pretty boys. I much prefer Happy Boy.

D: They're talented singers but not much to look at. They're just **rip-offs**② of the original Super Girl?

G: I've watched that since it started in 2004 but Super Girl itself is just a copy of American Idol.

D: Which was originally taken from Britain's Pop Idol! Who do you think is the best Super Girl?

G: I've always liked Li Yuchuan because of her boyish good looks and the fact that she sang songs previously sung by male singers which made her different from everybody else.

D: Speaking of boyish good looks what about Xu Fei with her short hair and shirt and tie?

G: It seems the androgynous look is a trend now but I much prefer girls to look like girls rather than tomboys!

D: And some of the boys look like girls, especially Xiang Ding! They are so pretty, emotional and feminine.

G: So now we have handsome women and pretty boys!

D: Too much gender confusion for me! I have to admit that the girls look cute but the boys look gay.

G: So you're never going to be a pretty boy!

习惯用语 2

① pin-up show: 帅哥美女秀节目
② rip-offs: 副本，拷贝

D: 你觉得"加油好男儿"这个节目怎么样？

G: 这节目不过是帅哥美女秀罢了，就是给女孩子们看那些头脑简单的"花瓶"男生。我还是更喜欢"快乐男生"。

D: 快男确实都很有唱歌天赋，但也没什么可看的，不过是个"超级女生"的翻版。

G: 自从 2004 年节目一开始我就一直在看。不过"超级女生"也只是"美国偶像"节目的翻版。

D: 其实这类节目的原创是"英国大众偶像"节目。你觉得最棒的超女是谁？

G: 我喜欢李宇春，她男孩气的外表，还有她所唱的那些以男生为原唱的歌曲，都使她独具特色。

D: 说到男孩般的外表，许飞不也是嘛，短发、衬衫，还有领带。

G: 现在，中性的外表似乎是一种潮流啊。不过我还是更喜欢端庄秀丽的女孩，而不是假小子。

D: 有些男孩看上去像女孩，尤其是"加油好男儿"里的项鼎！他们如此的漂亮、细腻感性和阴柔。

G: 是啊，现在就是英俊的女孩和漂亮的男孩的时代。

D: 我简直都搞不清性别了！那样的话，我觉得女孩看上去比较可爱，而男孩子看上去都像是同性恋。

G: 所以你永远不会想成为漂亮的男生吧。

Questions 2

1. Who's your favourite talent star? Why?
2. Which is your favourite talent show? Why?

Dialogue 3

Lili: Have you heard of Eason Chan? I think he's just **to die for**[①], so handsome!

Mike: Is that all you girls care about? Good looks? Whatever happened to talent, I bet he can't even sing.

L: Nonsense, he has the voice of an angel. He's very talented and even writes some of his own songs.

M: I think music has changed a lot in recent years, all over the world. For example, when I was at university, we didn't like singers because of their looks or how popular their music was. We loved them for being real.

L: What do you mean by "real"? Everyone is real, see touch my arm, you can feel me, so I'm real.

M: That's not what I mean by real, I mean really talented, real musicians.

L: But why isn't someone like Eason a real musician? He can sing very well.

M: You have lots of his CDs, right? You've seen him perform on TV, right? But have you ever seen him perform live?

L: I went to his concert last year and it was really him singing on the stage. He sang lots of his songs and danced and….

M: Yes, but did he really sing or did he mime? Some singers need computers to make their voices sound better, so when they do a concert they mime to the words. They don't really sing.

L: I see. Well, I don't know about that, but I think he sounded amazing. He's a wonderful entertainer. I'm really there for the atmosphere anyway, there is something special about being around so many people who all enjoy the same music.

M: Many famous people mime nowadays all over the world. But I totally agree with you about the amazing, electric atmosphere of a concert. Next time you go to a concert I'll come along, I want to see more of this guy. Can he really sing or is it just because girls think he's lovely?

习惯用语 3

① to die for: 渴望

L: 你听说过陈奕迅吗？我觉得他简直帅死了！

M: 你们女生只关心这些啊？仅仅是好的外表？天赋如何？我打赌他唱歌不怎么样。

L: 胡说！他的声音简直如天使一般。他极具天赋，还自己创作歌曲呢。

M: 我觉得近几年全世界的音乐发生了很大的变化。比如在我上大学的时候，我们不会因为一个歌手的外形或者他的歌曲有多流行而喜欢他。我们喜欢一个歌手，是因为他们的真实。

L: 你说的"真实"是什么意思？每个人都是真实的，你能够摸到我的胳膊，所以我也是真实的。

M: 我说的不是那个意思。我是说真正的天赋，真正的音乐家。

L: 那你为什么就说像陈奕迅这样的就不是真正的音乐家呢？他唱歌真的很棒。

M: 你有不少他的 CD，对吧？你看到过他在电视上表演，对吧？可你看过他的现场演出吗？

L: 我去年看过他的演唱会，就是他真真正正地在舞台上唱歌啊。他唱了好多首歌，而且还跳舞……

M: 他是真唱还是假唱呢？有些歌手需要电脑合成，才能使他们的声音变得好听。所以这种歌手在开演唱会的时候，只是对口型罢了，不是真正的唱歌。

L: 我明白了。这一点我也不敢肯定，不过我觉得他的歌儿听起来真的很棒。他是个很棒的歌手，那种气氛很让人沉醉，和一群他的粉丝在一起，感觉真得很特别。

M: 现在世界上的很多名人都假唱。不过，说到演唱会令人震撼的现场气氛，这点我是完全赞同你的。下次你再去看的时候，我也一起去。我想多了解一些这个歌手，看他到底是个真正的实力歌手，还是仅仅靠女孩子们对他的喜爱。

Questions 3

1. Who's your favourite male singer?
2. Who do you think is the most handsome?

15 艺 术
The Arts

‖ Background Information ‖

The arts can include painting, sculpture, literature, music, pottery, crafts, plays and many other things as well. They can be traditional or modern or a combination of both. The arts are the cultural heritage of any country and often reflect a golden age in that country's history.

艺术包括绘画、雕刻、文学、音乐、诗歌、手工艺、戏剧以及很多其他的艺术形式。艺术可以是传统或现代的，或两者兼而有之。艺术是一个国家的文化遗产，它常常反映了那个国家历史上的黄金时代。

背景信息

Dialogue 1

Melissa: This picture we see here is of a very traditional Chinese dragon.

Nick: It doesn't look like a Western dragon at all.

M: The Chinese dragon is said to have the head of a camel, the horns of a deer, the eyes of a rabbit, the ears of a cow, the neck of a snake, the belly of a frog, the scales of a carp, the claws of a hawk and the palm of a tiger.

N: That's nine different animals rolled into one!

M: Did you know we also connect dragons with the number 9? And 9 for us is a very lucky number.

N: I didn't know that. For us dragons have small wings, can fly, breathe out fire, hoard treasure, are fierce and evil in nature.

M: Well, in China dragons are not seen **in the same way**[①] as in England for example. They are not evil, but lucky.

N: Lucky, in what way?

M: They are seen as being benevolent. The Chinese people are sometimes called the Offspring of the Dragon so dragon culture is very popular.

N: So they are very important to Chinese people, right?

M: Yes, although we don't know the exact origins, we are sure they are very ancient. We can find them in literature, poems, songs and architecture.

习惯用语 1

① in the same way: 同样地

M: 咱们现在看到的这个图片上画的是一条传统的中国龙。

N: 看上去和西方的龙完全不同。

M: 据说，中国的龙拥有骆驼的头、鹿的角、兔子的眼睛、牛的耳朵、蛇的颈部、青蛙的腹部、鲤鱼的鳞片、鹰的爪、虎的手掌。

N: 那可是九种不同的动物合为一体啊！

M: 你知道嘛，我们经常把龙与数字"9"联系在一起。"9"对我们来说是个非常幸运的数字。

N: 这个我倒不知道。我们西方的龙有一对小翅膀，可以飞翔、喷火、藏匿宝藏，天性凶暴邪恶。

M: 在中国，人们对于龙的看法可与英国不同。把它看作是吉祥而不是邪恶的化身。

N: 吉祥？为什么？

M: 中国龙是慈善的代表。中国人有时称自己为"龙的传人"，所以，龙文化非常盛行。

N: 所以，龙对于中国人来说非常重要，对吗？

M: 没错，尽管我们不知道龙的确切起源，但我们确信很早以前就有了。在文学巨著、诗歌、歌曲以及建筑物中都可以找到龙的影子。

Questions 1

1. Do you have any dragon pictures or motifs in your apartment?
2. Why is the number 9 lucky?

Dialogue 2

Nick: Everybody keeps asking me what my animal sign is. Why is that?

Melissa: It's part of our culture. The twelve animals have different characteristics and they apply to the people born in that year. However, they also apply to the hours of the day and some people think that the time of day when you were born is more important than the year.

N: But there are twenty four hours in one day!

M: Right, so every two hour period is connected to an animal.

N: We have the twelve signs of the zodiac so to us it's the month

Culture 文化

that is important not the year or the hour.

M: Another reason why people ask you for your animal sign is that they can work out how old you are.

N: That's cunning of them! Because the signs are every twelve years then for example if I was born in the year of the pig then I would be either twenty four or thirty six, right?

M: Or forty eight!

N: You cheeky monkey!①

M: Actually I'm a tiger! So you'd better be careful or else I'll eat you!

N: So where did the animals come from?

M: There's a story that Buddha invited all the animals to his kingdom but only twelve turned up. Out of gratitude Buddha decided to name the year after each of the animals in order of their appearance.

习惯用语 2

① You cheeky monkey! 你这只皮猴子！（用来形容某人的无礼）

N: 人们为什么总是问我的属相是什么啊？

M: 这是中国文化的一部分。十二生肖各有特征，生于那一年的人就是那个属相。同时，十二生肖也用于命名一天中的不同时间，有些人认为，生于一天当中的几时，比生于哪一年更重要。

N: 可是一天有 24 个小时啊！

M: 没错，所以每两个小时被划分为一种时辰。

N: 我们西方拥有十二星座的标记，所以，对于我们来说，出生的月份比年份或者具体时间更重要。

M: 人们问你属相的另一个原因就是想计算出你的年龄。

N: 简直太狡猾了！十二个生肖为一轮，所以要是我说自己属猪的话，那我要么是 24 岁，要么是 36 岁，对吗？

M: 也可能是 48 岁！

N: 你这只皮猴子！

M: 其实我属虎！你最好小心点儿，不然我就吃了你！

N: 这些动物是怎么来的？

M: 相传，佛祖邀请所有的动物到他的王国去，但最终只有这 12 种动物去了。为了感谢他们的出席，佛祖决定用他们的名字命名年份，并按照当时它们到场的顺序来命名每一年。

Questions 2

1. Do you think your animal sign has any bearing on your character or future?

2. Do you ever ask for someone's zodiac sign so you can work out their age?

Dialogue 3

Nick: You know, I'm thinking of doing something different. I always seem to go jogging to exercise and I want to try something new, but something traditional. Any ideas?

Melissa: Why not try Qi Gong. My grandfather is an excellent teacher and it will give you a chance to practice your Chinese, too.

N: I don't know Qi Gong? Isn't it very violent? I don't want to learn to fight!

M: That's a common misconception, Qi Gong is an art, and it's slow and beautiful. The movements are like running water.

N: OK. Why do people practice this? I guess it must have a long history if it has been elevated to the level of an art.

M: Yes, Chinese people have been practicing it for thousands of years. Its main purpose is to cultivate the Qi we all have inside and learn to control it. My grandfather says that when he practices, he can feel his Qi moving around his body.

N: I read somewhere that if you can harness the power of your Qi, you can shoot it out of your body to another, to either harm or heal. Is that true? I don't want to be a bad student and your grandfather gets so angry with me that he makes a killer move!

M: Haha! Don't worry, he's a patient old man, but he could kill you with one finger! Just kidding. Seriously, I don't know what's true. I've read the same things as you and seen it in many movies being used to kill people, but I'm not sure.

N: So you believe it's just a good way to stay in shape?

M: I think so, yes. When you see people practicing they just look so natural, yet so precise. It's a strange mixture, I guess. It's good for everyone too, from little kids up to old people.

N: OK, I think I'll try. If an old guy can do it, surely a strong 40 year old like me can. When shall we go and ask your granddad?

M: Oh, I forgot to tell you. We will have to go there in the morning, at perhaps 5 am. He practices for his morning exercises, you see. I hope you're a morning person, because if you want him to teach you you're going to have to be **up at the crack of dawn**①!

习惯用语 3

① up at the crack of dawn: 很早就醒来

N: 你知道，我每天都是慢跑来锻炼，太没意思了，所以我想做些不同的事儿。我想尝试些既新鲜又传统的东西。你有什么好主意吗？

M: 你可以试试气功。我爷爷就是个很棒的气功师傅，而且你也可以顺便练练中文。

N: 什么？气功？那不是很暴力吗？我可不想学打架。

M: 人们通常都有这样的误解，其实不是，气功是一种艺术，动作缓慢而漂亮，如行云流水一般。

N: 好吧，人们干嘛练这个？我想既然称之为艺术，那它一定有很长的历史了吧。

M: 是的，中国人练气功已经有几千年的历史了。其主要目的就是学会培养并控制我们体内的气。我爷爷说，当他练气功的时候，可以感到一股气在体内游动。

N: 我读过一篇文章，要是你能够掌握体内气的能量，你就可以将其注入另一个人的体内，或伤及他人，或治疗疾病。这是真的吗？我不想当个坏学生，要是惹你爷爷生气了，他把我杀了怎么办！

M: 哈哈！别担心，他很有耐心的。不过他也可以用一个手指头就杀了你！开个玩笑。不过说实话，我也不知道这是不是真的。我也读到过类似的东西，而且在很多电影里也看到过用气杀人的场景，但我不知道是真是假。

N: 你确定这一定是一种保持体型的好方法？

M: 嗯，没错。人们在练气功的时候表情很自然，而且有板有眼的，真是不可思议。无论是小孩还是老人，练气功都有一定的益处。

N: 好吧，那我试试。要是老人可以，那像我这样 40 来岁的中壮年也一定可以。咱们什么时候去找你的爷爷请教？

M: 哦，忘了跟你说了。咱们得早上去，大概 5 点钟。他晨练的时候才练气功。但愿你不是个大懒虫，要想跟他学气功，你就得早点儿起床。

Questions 3

1. Have you ever tried Qi Gong?
2. Which kind of exercises do you like to do?

Movies

16　电 影

‖ **Background Information** ‖

The 1980s and 1990s was the golden era but also the most commercially exploited era of Chinese movie making. There was a flood of Chinese movies ranging from martial arts, to love stories, to slapstick comedies, to Hollywood copycats.

There were some commercial success but most movie ventures lost money. Breaking the trend of Chinese movie flops was Jacky Chan, an up and coming martial art actor who laced his kung-fu movies with a huge dose of humour and sometimes slapstick. Around the same period, Tsui Hark, another great Hong Kong director, directed and produced a few blockbusters around the legends of an early Kung Fu master, Huang Feihong. He also brought to fame Jet Li, a top martial arts champion from China, who was the key actor for many of his movies.

Not only did many of these Chinese movies take Chinese speaking society by storm, they also took Asia and eventually the rest of the world by storm. Chinese movie making talent had arrived on the world stage.

上世纪80、90年代，是中国电影业发展的黄金时期，也是电影业最商业化发展的时代。当时涌现出大量的中国电影，包括功夫片、爱情片、喜剧片，以及类似好莱坞的大片。

有些电影取得了商业上的成功，但是大多数都亏本。带领中国电影业走出低谷的是成龙，这位功夫演员在中国的功夫电影中融进了很多幽默和搞笑元素。在这个时期，另外一位香港名导演徐克，围绕着中国功夫大师黄飞鸿的生平事迹，导演并制作了好几部功夫大片。这也捧红了功夫明星李连杰，他本来是中国武术锦标赛的冠军，后来他在徐克导演的众多影片中扮演主角。

这些中国电影不仅轰动了当时的华语社会，而且也轰动了亚洲和世界各地。中国的电影业已经迈向国际舞台。

背景信息

Dialogue 1

Tori: Hello. I'm looking for Zhang Yimou films. I just saw *Hero* and think he is a great director.

Lana: Yes, Zhang Yimou has become very popular in recent years. His recurrent theme is a celebration of the resilience, even the stubbornness, of Chinese people in face of hardships and adversities. He is also well known for his sensitivity to colour. The first movie he directed was *Red Sorghum* in 1987 which won the Golden Bear at the Berlin Film Festival. The movie is full of dynamic edits, striking **close-ups**①, and gorgeously photographed images.

T: Just like *Hero*, he seems to always make beautiful movies.

L: Then after a few years he made the amazing *Raise the Red Lantern* starring Gong Li.

T: Yes, I've heard of this movie. I will buy that one if you have it.

L: Here it is. Also, in 1992 he made this one, *The Story of Qiu Ju* about a pregnant lady whose husband has been beaten up by a village leader. This movie is very different from his others, there is no rigid style or sumptuous photography, it was filmed in a documentary style, which gives it a **gritty look**②.

T: That one's very different, I'll take it too.

L: As for more recent films, in 1999 *Not One Less* won the coveted Golden Lion at the Venice Film Festival.

T: What about movies from the 2000's?

L: In 2002 he made *Happy Times* which was a **seriocomic drama**③, then, of course *Hero*. And in 2004 *House of Flying Daggers*, which was set in the Tang dynasty and had a flamboyant use of colour, very beautiful.

T: And most recently?

L: In 2006 he made *Curse of the Golden Flower*, an amazing **period epic**④. It stars Gong Li, Jay Chow and Chow Yun

Culture 文化

Fat.

T: What's he working on now?

L: The opening ceremony for the 2008 Olympics. If it's anything like his films it's bound to be a smash hit!

T: 你好，我正在找张艺谋的电影。我刚刚看了他拍的《英雄》，觉得他真是个非常不错的导演。

L: 没错，最近几年张艺谋非常走红。他作品的主题所展现的往往是中国人民在面对困难与不幸时所表现出的乐观而顽强的精神。另外他也因拍电影时对色彩的精湛运用而闻名。他第一部导演的作品是 1987 年拍摄的《红高粱》，这部电影获得了柏林电影节的金熊奖。该电影里有大量的动态剪辑、特写镜头和精美的拍摄画面。

T: 就像《英雄》一样，他总是能拍出漂亮的电影。

L: 在拍完《红高粱》之后，过了几年，他又拍了《大红灯笼高高挂》，由巩俐主演的。

T: 我听说过那部电影。你这里有吗？我要买。

L: 在这儿呢。另外，他在 1992 年还拍摄了《秋菊打官司》，讲的是一个怀孕的妇女，她的丈夫被村长打了。这部电影不同于他的其他作品，没有延用他惯用的拍摄模式和华丽镜头，而是采用纪实的手法来拍摄的，给人一种真实感。

T: 那还真是很特别，我买。

L: 至于最近的作品就是 1999 年的《一个都不能少》，获得了威尼斯电影节的金狮奖。

T: 2000 年以后有什么新电影吗？

L: 2002 年，他导演的《欢乐时光》，是一部悲喜交加的电影。然后就是《英雄》。2004 年的《十面埋伏》，是一个发生在唐朝的故事，影片色彩华丽，非常漂亮。

T: 最近的呢？

L: 2006 年拍摄的《满城尽带黄金甲》是一部非常好的历史戏。由巩俐、周杰伦、周润发主演。

T: 他现在在拍些什么？

L: 他在准备 2008 年奥运会的开幕式。要是能和他的电影一样，那肯定也非常轰动。

Questions 1

1. Who is your favourite director? Why?
2. What do you think the opening ceremony for the 2008 Olympics will be like?

Dialogue 2

Kate: Which Chinese actress do you like the best?

Tony: I think I'd have to say Gong Li. Ever since she starred in *Red Sorghum* and *Raise the Red Lantern* she's been China's number one.

K: For over twenty years! I much prefer Zhang Ziyi. I thought she was wonderful in *Hero*.

T: She was I agree but in *Memoirs of a Geisha*, although her acting was good, I felt she was **let down**① by her poor English. Who do you think's going to be the next Gong Li?

K: Everyone says it must be Zhang Ziyi.

T: Well, I've just seen *Protégé* so I've got to say it's Zhang Jingchu. And unlike Zhang Ziyi her English is really fluent.

K: Yeah English is **a must**② for Chinese actors now as they all seem to end up in Hollywood.

T: That's where Zhang Jingchu made her latest film *Rush Hour 3*. It's **due out**③ in August.

K: I know very little about her. What other films has she been in?

T: Her first film was *Peacock* in 2005 so she's definitely the rising star of this millennium. All she needs is a male costar. Wouldn't

Culture 文化

it be great to have a Chinese couple rival Brad Pitt and Angelina Jolie! Any suggestions?

K: I can't think of anyone Chinese. What about Rain? Now they would make a handsome couple!

T: Yeah. A match made in Hollywood heaven!

K: Now that would definitely be a blockbuster! And who knows, they may even **hit it off**④ and then we could see the rise of an Asian film dynasty.

T: Yeah. Forget Hollywood. Forget Bollywood. Here comes Chollywood!

习惯用语 2

① let down: 失望   ③ due out: 公映日期
② a must: 必不可少的东西   ④ hit it off: 合得来，相处融洽

K: 你最喜欢哪位中国女演员？

T: 巩俐。自从出演《红高粱》和《大红灯笼高高挂》开始，她就奠定了自己中国女一号的位置。

K: 而且一直持续了二十多年！不过，我更喜欢章子怡。我觉得她在《英雄》里的表演简直精彩极了！

T: 没错，我同意。可是在《艺妓回忆录》中，尽管她的表演也十分精彩，但她那糟糕的英语真是令人失望。你觉得谁会成为下一个巩俐？

K: 肯定是章子怡，大家都这么说。

T: 我刚刚看了《门徒》，所以我觉得会是张静初。她的英语相当流利，不像章子怡那样。

K: 没错，对于中国演员来说，他们最终想要闯进好莱坞，英语是必需的。

T: 张静初最近在好莱坞拍摄了《尖峰时刻3》。8月份就会公映。

K: 我对张静初了解很少。她还演过什么别的电影吗？

T: 她的第一部电影是2005年拍摄的《孔雀》，很显然，她是一颗冉冉升起的新星。她需要个男搭档。要是有一对中国搭档可以与布拉德·皮特和安吉丽娜·朱莉相匹敌，那多棒啊！你有什么好的人选吗？

K: 我实在想不出有哪个中国男演员适合。Rain 怎么样？绝对是一对帅气的组合！

T: 嗯，这绝对是"好莱坞造好莱坞设"的一对！

K: 那绝对是一鸣惊人啊！也许他们是绝配，从而让我们看到亚洲电影时代的到来呢！

T: 是啊！到时候，什么好莱坞，什么宝莱坞，统统滚蛋，"华莱坞"时代即将来临了！

Questions 2

1. Who's your favourite actress? Why?
2. Who will be the next Gong Li?

Dialogue 3

Paula: Which Chinese actor do you like best?

Nick: I think the first Chinese actor I ever saw was Bruce Lee. I really loved his Kung Fu fighting.

P: But Kung Fu has moved on since then. There's Jet Li and Stephen Chow for example.

N: I really like Jackie Chan. I've seen most of his movies and I enjoyed his latest *Rush Hour 3*.

P: But Jackie Chan is **getting past it**① now and I've heard he's looking for his successor now.

N: The only other actor I know is Chow Yun Fat and he's getting old too.

P: You need to keep up with the times. The Chinese movie industry has moved on since the old Hong Kong days.

N: Yeah, I really liked them. They were very funny and the English subtitles were hilarious because of the bad English.

P: There are many new and up-coming directors now. Some of them bypass the traditional studios and use hand-held cameras or even mobile phones.

Culture 文化

N: But the quality can't be much good.

P: It's all about using new technology and also doing things in a different way. Using mobile phones forces you to do lots of close-ups and have a strong story line.

N: So I'll be able to watch them on my MP4 or my iPhone!

习惯用语 3

① get past it:（因年老等）精力不济的，无法做年轻时能做的事的

P: 你最喜欢哪位中国男演员?

N: 我最早看到的中国男演员是李小龙。我很喜欢他的功夫电影。

P: 不过，中国功夫从那以后已经有了很大的发展，比如现在的代表人物有李连杰和周星驰。

N: 我很喜欢成龙。我看过很多他的电影，他的新片《尖峰时刻3》很不错。

P: 但是成龙已经老了，我听说他正在寻找自己的继承人。

N: 那我所知道的男演员里就只剩下周润发了，不过他也老了。

P: 你得与时俱进啊! 在那些老的香港电影的基础上，中国电影业已经有了很大的发展。

N: 香港的那些老电影我很喜欢，很有趣，尤其是看了那些蹩脚的英文字幕，就更搞笑了。

P: 现在也涌现出了很多新锐导演。他们走出传统的摄影棚，用手提摄像机甚至是手机进行拍摄。

N: 那画面质量肯定不太好。

P: 这就是使用新技术，另辟蹊径。用手机拍电影得需要你多拍特写镜头，而且还要有很强的故事情节。

N: 那我就可以在手机或者 MP4 上看电影了!

Questions 3

1. Who is your favourite male star? Why?
2. Which was the last movie you saw?

17 公共交通
Public Transport

‖ Background Information ‖

In cities the main forms of public transport are buses, taxis and the subway.

There are more than 20,000 buses in Beijing. Half of them use natural petroleum gas. There are 648 bus routes in Beijing that transport 10 million people each day. Beijing plans to open 50 new bus routes every year. In 2003, Beijing is expected to have 650 bus routes with annual passengers of 4.5 billion. Besides, there are 222 long-distance bus routes linking downtown area with suburban districts and surrounding regions. At present, there are 67,000 taxis in Beijing. By the end of 2008, all the taxis will be equipped with a wireless telecommunication system and Global Positioning System (GPS).

在城市里，主要的公共交通工具是公交车、出租车和地铁。

在北京有2万多辆公交车，其中一半使用天然气作燃料。有648条公交线路，每天运送乘客1000万人次。北京计划每年增开50条新的公交线路。在2003年，北京已有650条巴士线路，年运输量达45亿人次。此外，有222条长途汽车线路，连接着郊区及远郊区县。目前，北京有出租车6.7万辆。到2008年底，所有出租车都会安装无线通讯系统和全球卫星定位系统。

背景信息

Dialogue 1

Nick: This bus is so crowded! Oh good it's stopping to let some passengers off.

Joanne: We'll be able to breathe a bit now. It's so squashed in here.

N: Oh no! He's letting more passengers on! Is it always like this?

J: It is rush hour you know so we shouldn't complain. Anyway we'll be at our stop soon.

N: Thank heavens for that! I don't think I can stand it any more.

J: What are buses like in Britain?

N: We have a lot of single deckers like this but the number of standing passengers is strictly limited.

J: But you don't have as many passengers as we do.

N: Good job as I don't know how we would cope!

J: I know many people in Beijing want to buy cars because they think the buses are too overcrowded, dirty and slow.

N: But that would mean even more cars and the rush hour would get even worse! I've heard that rush hour in the CBD starts at 7 am and lasts until 8 pm!

J: Perhaps the government should ban any more new cars and start to extend the subway as much as possible.

N: They need to do something quick or else **the gridlock**[①] will spread all over the city!

习惯用语 1

① the gridlock: 因堵塞而使交通瘫痪

N: 这车可真挤啊! 好在到站了，下去了一些人。

J: 现在可以喘口气了，这儿太拥挤了。

N: 天哪! 上来的人比下去的还多! 这车总是这样吗?

J: 现在是高峰时期，就别抱怨了。反正咱们马上就要到站了。

N: 谢天谢地！我都快受不了了！

J: 英国的公交车是什么样的情况？

N: 我们也有很多这样的单层巴士，不过站着没有座位的乘客数量是受严格限制的。

J: 但你们的乘客也不像我们这么多啊。

N: 幸好这不是在英国，因为这要是在英国，我还真不知道该如何解决。

J: 我知道因为公交车又挤、又脏、又慢，所以很多在北京生活的人都想买私家车了。

N: 这不就意味着车越来越多，高峰时的路况越来越糟糕吗？我听说北京 CBD 地区的拥堵时间从早上 7 点一直持续到晚上 8 点！

J: 我想政府应当禁止新增机动车，而且要尽可能发展地下交通。

N: 政府得赶快采取点儿措施，不然，堵车状况会蔓延到整个京城的！

Questions 1

1. What's the best thing about travelling by bus?
2. The worst thing?

Dialogue 2

Kevin: My goodness! It's getting too crowded, I'm starting to feel claustrophobic.

Audrey: What does that mean?

K: It means that there are too many people around me and I feel like everything is closing in on me. I'll never get used to the subway at rush hour.

A: Don't worry, only two more stops, then we are at the interchange station. Lots of people will get off there.

K: I hope you are right. I don't get how you can actually enjoy taking the subway! I mean, don't you feel like people are interfering in your **personal space**①.

A: I guess I'm just used to it now. I've been taking the subway to work for years, it's cheap and convenient. If I take a bus, it's just as crowded and I have to change buses three times.

K: On the subway we only need to change from line 13 to line 2, so it is pretty easy. But on the bus you can see out of the window. Down here you can't see anything interesting.

A: You can't always see out of the window on the bus, especially if it's as packed as this.

K: Maybe I will never be able to get used to the subway. We don't have one in my hometown, you see.

A: Didn't you ever take the tube in London? I hear that London's underground is the best in the world.

K: I took it a few times, when I was there visiting, because the taxis are too expensive. I never liked it then, either.

A: Well, if you are going to continue living here cheaply, you'll have to get used to it. You should be happy we don't live in Tokyo, then you'd really know what crowded is!

习惯用语 2

① personal space: 私人空间

K: 我的天哪! 太挤了! 我都快犯幽闭恐怖症了!

A: 什么是幽闭恐怖症?

K: 就是说我周围有太多的人, 让我觉得似乎是被封闭住了。我永远都无法习惯高峰时期的地铁。

A: 别着急, 还有两站就到换乘站了, 那站下车的人多。

K: 但愿你是对的。我真不明白你为什么喜欢坐地铁。我是说, 你不觉得别人离你太近了, 干涉了你的私人空间吗?

A: 我坐地铁上班已经好几年了, 现在已经习惯了, 坐地铁便宜而且方便。要是坐公交车, 不但跟现在一样拥挤, 而且我还得换三次车。

K: 要是坐地铁，我们只需要从 13 号线换到 2 号线，这样简便多了，但是坐公共汽车，你可以看看窗外的景色，坐地铁什么有趣的东西都看不到。

A: 坐公交车也不是总能看到窗外啊，尤其是像现在这样拥挤的时候。

K: 也许我永远也无法适应地铁。我的家乡没有地铁。

A: 你在伦敦没坐过地铁吗？我听说伦敦的地下交通是世界上最发达的。

K: 去那儿旅游的时候坐过几次，因为出租车实在是太贵了。伦敦的地铁我也不喜欢。

A: 你要是想继续在这里以低成本生活的话，就得习惯地铁。幸亏我们不是住在东京，到了那儿你才知道什么是真正的拥挤。

Questions 2

1. What do you like best about the subway in your city?
2. What changes would you make?

Dialogue 3

Nick: So, Ann. What's your favourite way of getting around in Beijing?

Ann: To be honest, I wish I had my own car. It would be great to have the freedom to go anywhere I wanted at any time I liked.

N: You're joking, aren't you? There are already far too many private cars in Beijing, do you want to add to the pollution?

A: OK, by taxi then. But it's too expensive. I used to bike everywhere, but nowadays I usually take the subway. It's a little limited, though.

N: What do you mean by limited? You can go to lots of famous places by subway, such as Tian'anmen, Wangfujing, Xidan.

Travel 旅行

A: Yes, but there are lots of other places you can't get to. Take my new apartment, for example. The government has tried to encourage us to live outside of the city, but the transport links are too bad.

N: So there's no subway up there on the 5th Ring Road?

A: There is one, but to get to it I must either walk for 40 minutes on a very rough road or take an illegal taxi.

N: I see. What about buses? Couldn't you take a bus to work?

A: To get to downtown, during rush hour it will take around 2 hours by bus. Luckily, they are building many new subway lines linking lots of Beijing. They should open in 2008.

N: I guess they're needed for the Olympics because of all the extra people that will flood into Beijing. And it will be great for you when you are going to work. I'm lucky, because I live at the university and teach at the university, I can just stumble out of bed 5 minutes before class!

A: Yes, you lucky devil. Oh well, I'll just have to be patient, 2008 is **just around the corner**[①], after all.

习惯用语 3

① just around the corner: 即将来到的

N: Ann，你最喜欢以什么方式逛北京城？

A: 老实说，我想有一辆属于自己的车。随时随地，想逛就逛，多好啊！

N: 你在开玩笑吧？北京的私家车已经够多的了，你还想再增加污染？

A: 好吧，那就打车，可是太贵了，我过去老是骑车，但现在改坐地铁了，尽管有些受限制。

N: 什么叫受限制？坐地铁可以到很多有名的地方去啊，比如天安门、王府井、西单。

A: 对，可也有很多地方到不了，就比如我的新家。政府一直鼓励我们住在城外，可交通路线实在是太差了。

N: 你是说五环路上没有地铁？

A: 有一条。不过要想到那个地铁站，我得要么在一条坑坑洼洼的路上走 40 分钟，要么打一辆"黑车"。

N: 我明白了。那公交车呢？你不能坐公交车去上班吗？

A: 高峰时期要想坐车到市中心，起码需要两个小时左右的时间。好在许多新的地铁线路正在建设当中，到 2008 年就可以投入使用了。

N: 奥运期间，大量游客涌入北京，这些地铁就有用武之地了。到时你上班也就方便多了。我很幸运，我住在大学里，又在大学里教书，只要在上课前 5 分钟起床就可以了。

A: 没错，你简直幸运死了！看来我只有耐心等待，毕竟 2008 年马上就要到了。

Questions 3

1. What's the best way to get about in your city?
2. Would you rather live centrally or in the suburbs?

18 汽车

Cars

‖ Background Information ‖

Nowadays, China is no longer the bicycle kingdom, as more and more people realize their dreams and buy their first car. According to a report by the Chinese auto industry association, China has surged past Japan to become the world's second largest vehicle market after the United States. By the end of May 2007, Beijing alone had 3 million automobiles crawling around the city's avenues, often stuck in heavy traffic. The capital currently adds around 1,000 cars to its total every day. In Shanghai, bicycles were banned all together from larger avenues as far back as 2004: an acknowledgement of the growing weight of the car in China. But what cars do Chinese people really want? China is expected to emerge as the largest luxury car market in the world within the next decade—which is a reflection of changing attitudes towards wealth.

现如今，随着大多数人实现了自己的梦想，买了自己的第一辆汽车，中国不再是自行车的王国了。中国汽车工业协会的一份报告显示，中国已经超越日本，成为了仅次于美国的世界第二大汽车市场。截止到 2007 年 5 月，仅北京每天就有 300 万辆机动车行驶于大街小巷，这常常会导致严重的交通问题。目前，还在以每天 1000 辆的速度增加。在上海，早在 2004 年就已经禁止自行车在主要大街上行驶，这同样表明了汽车在中国的份量。但是，中国人到底喜欢什么样的车？在未来 10 年内，中国有望成为世界上最大的豪华车市场——这反映了人们对于财富观念的转变。

背景信息

Dialogue 1

Bruce: Wow! Look at that cool car over there.

Tori: Which one? The QQ? I think they are such a cute, little car. But obviously only for girls, though.

B: Not the QQ, they are so cheap, the Buick. I think it's an imported one, the steering wheel is on the other side.

T: What's so special about a Buick? Now, if it were a Mercedes Benz or a BMW, I would agree with you. But a Buick? They are just a regular American car.

B: But that's the point, they are American and an imported one is very rare in China. Mercedes are only for bosses of government companies, older, boring men.

T: So you think Buicks are for younger, **funkier**① men?

B: Funkier? What does that mean?

T: You know, more fashionable, more stylish, but also a little different.

B: Exactly! How do you say that in English, you **hit the nail on the head**②. Most younger guys in China nowadays dream of having an imported car.

T: But aren't Mercedes imported? I mean, they are a German car.

B: Yes, they are a German car, but today they are also built in China. Young, rich men would not waste their money on a domestic car.

T: I see, so an imported car is more of a **status symbol**③. Things sure are different here!

Travel 旅行

习惯用语 1

① funkier: 时髦独特的
② hit the nail on the head: 说得好，正中要害
③ status symbol: 社会地位的象征

B: 哇！快看那辆车，多酷啊！

T: 哪辆？那个QQ？我觉得这款车很小巧，很可爱。但显然，它只适合女生。

B: 不是那辆QQ，那太便宜，是别克。我估计那是辆进口车，它的方向盘在另一侧。

T: 一辆别克有什么特别的？要是辆奔驰或者宝马，我还能理解你为什么那么吃惊。可是别克？不过是辆普通的美国车罢了。

B: 就是因为这个啊，在中国，美国的进口车可不多见。奔驰都是给那些政府和公司里的大老板、无聊老男人们坐的。

T: 所以你觉得别克适合那些年轻、明髦有个性的年轻男士？

B: 明髦有个性？什么意思？

T: 就是更流行，更时尚，但又与众不同。

B: 没错！用英语怎么说？你说得没错，如今，中国的年轻人都梦想有一辆进口车。

T: 奔驰不就是进口的吗？我是说，奔驰是德国车。

B: 没错，是德国车。但现在在中国也生产，有钱的年轻人是不会把钱浪费在国产车上的。

T: 我明白了，进口车更是一种社会地位的象征。这里人们的观念可真不同啊。

Questions 1

1. If you could buy any car, what would it be?
2. Do you think what you drive, shows who you are?

Dialogue 2

Daisy: Tori, what do you think about colours when selecting a car?

Tori: Do you mean what colour would I select? Well, I think I would have to go with yellow. Yellow is bright and **summery**①. Or maybe green.

D: Green! That's funny, do you want a new job as a post

woman?

T: What do you mean? I just like the colour, that's all. I think green is quite a popular colour for cars, especially the dark, shiny green called "British Racing Green".

D: Really? But in China, when we see a dark green car we think of the China postal service. The postal workers who deliver mail and parcels all drive small, dark green vans.

T: So, what colours are popular in China? I've seen lots of white cars, but to be honest I think most Westerners don't like white cars because we have to clean them too often.

D: Well, I think most people would like to drive a black or a red car.

T: Why are those colours so popular?

D: Red means happiness and power, black means power and high position.

T: So I guess Chinese people think that power is important. In Western countries, we think that red means anger or passion. Have you ever heard the phrase "**I'm seeing red**②"?

D: No, I haven't. But, you mean that if you get furious, your eyes will fill with blood?

T: Sort of, yes. When we get too angry, all the blood rushes to our heads, our faces change colour, so we see red.

D: We have a saying that if you get too angry all of your hair stands on end and pushes your hat off. Our two sayings mean the same thing, but are very different.

习惯用语 2

① summery: 夏季的，如夏季的　　② I'm seeing red: 我很生气

D: Tori，选车的时候你会考虑什么颜色？

T: 你是说我会选择什么颜色？我选黄色，明亮而且会让人联想起夏

天。也可能会选绿色。

D: 绿色！太搞笑了，你想去当邮递员吗？

T: 什么意思？我只是喜欢绿色罢了，就这么简单。对于车来说，绿色很流行，尤其是那种深而发亮的绿颜色，也叫"英国赛车绿"。

D: 真的吗？可是在中国，我们一看到深绿色的车，就会想起中国邮政。那些负责递送邮件和包裹的邮政人员就开小型的、深绿色的货车。

T: 那在中国什么颜色流行？我看到有好多白色的车，不过说实话，我觉得大多数西方人不喜欢白色的，你老得清洗它。

D: 我想大多数人还是会喜欢开红色或黑色的车。

T: 这两种颜色为什么这么流行？

D: 红色意味着快乐与权力，黑色意味着权力与地位。

T: 我想中国人把权力看得很重要。在西方国家，我们认为红色代表着气愤或激情。你有没有听过这句话"I'm seeing red"？这表示"我非常生气"。

D: 没听过。你是说，如果激怒了你，你的双眼会充满血腥？

T: 差不多。当我们生气的时候，大量血液会涌到我们的头部，我们的面部就会变色。这样我们看起来就"很红"。

D: 我们经常这么说：要是你很生气，你的头发就会竖立起来，把你的帽子顶掉，就是"怒发冲冠"。咱俩说的是一个意思，只是表达方式不同。

Questions 2

1. What colour would you select if you bought a car?
2. Do you think colours have hidden meanings?

Dialogue 3

Adam: I heard you wanted to buy a new car. Did you get one yet? I have a friend who works in a car showroom, so maybe he can give you a discount.

Joe: Yes, I do want to get a car, but not a new one. I want to buy a **second-hand**^① one.

A: What do you mean by second-hand? A car doesn't have any hands!

J: No, "second-hand" means "used". Someone else has already owned it, maybe driven it for a few years then decided to change it.

A: You mean an old car! Most people in China like our cars to be new, we don't like the thought of somebody else owning it before us.

J: But why not? A second-hand car is much cheaper than a new one, and if it's in good condition, it's good value for money.

A: So, are second-hand cars very common in Western countries? In England, for example. English people are rich, so I'm sure they don't buy old cars.

J: Actually, buying second-hand cars is very popular in England, because the taxes on new cars are very high.

A: I'm surprised by that! Did you have a second-hand car when you were in England?

J: Indeed, maybe even third or fourth hand. I had a **vintage**^② MGB GT.

A: Vintage in English, can describe very old wine. So, you mean a very old car?

J: Not very old, around 25 years. It is a great hobby in England to collect vintage cars and drive them for fun at the weekend.

A: I'm not sure about that. I think you should still let me ask my friend to help you buy a new car!

Travel 旅行

A: 我听说你要买辆新车，买了吗？我有个做汽车销售的朋友，没准他能给你打个折。

J: 没错，我是想买辆车，但不是新车。我想买辆二手车。

A: "二手"是什么意思？汽车没有手啊。

J: 不是，"二手"的意思就是"别人用过的"。别人的、开了几年现在想卖掉的。

A: 你是说旧车！在中国，大多数人都希望自己的车是全新的，我们可不喜欢别人用过的东西。

J: 为什么啊？二手车比新车要便宜，而且要是车况好的话，性价比还是很高的。

A: 二手车在西方国家很普遍吗？比如在英国。英国人那么有钱，我保证他们都不会买旧车的。

J: 事实上，买二手车在英国很流行。因为新车的税非常高。

A: 太让我惊讶了！你在英国的时候买过二手车吗？

J: 买过啊，甚至是"三手"、"四手"的都有。我就有一辆老款的MGB GT 车（这是汽车生产商的名字，全称是 Morris Garages，而 GT 是 grand tourer——译者注）。

A: "Vintage"在英语里可以用来形容年代久的葡萄酒，如陈年佳酿。你这里说"vintage"，你的意思是指一辆古老的车？

J: 也不是很老，25 年左右。在英国，收集老式汽车，然后在周末的时候开着出去玩儿，是一大乐事。

A: 这我可说不好。我觉得你还是应该让我问问朋友，帮你买辆新车。

Questions 3

1. Would you ever buy something that was second-hand?
2. Do you think that collecting old cars is an interesting hobby?

19 自行车

Bicycles

Background Information

Nationally, there are 470 million bikes and 4 million of them go missing every year. A recent survey by a Renmin University student found that on average each student lost 2.3 bikes. The number of stolen bikes recovered by the Beijing police was 780 for April 2007.

全国共有 4.7 亿辆自行车，这其中，每年会有 4 百万辆自行车丢失。最近，一项对中国人民大学学生的调查显示，平均每个学生丢失过 2.3 辆自行车。2007 年 4 月，北京警方追回的被盗自行车有780 辆。

背景信息

Dialogue 1

Nick: There are so many bikes on the road!

Gloria: We used to be known as the kingdom of bicycles but actually there are not as many bikes on the road as there used to be.

N: In England we don't have so many bikes and we have very few bicycle lanes too.

G: Why is that?

N: Well, most children would have bikes but when they get old enough to drive they would usually buy a car.

G: Do British students have bikes?

N: If they do have a bike it's likely to be a **mountain bike**[①] but most students would have a car.

G: They must be very rich, then!

N: Not at all. They would buy a cheap second-hand car.

G: I heard that students in Cambridge and Oxford use bikes. Is that so?

N: Yes but it's more of a tradition than anything else although Cambridge and Oxford are both small cities and the colleges are in the middle of the city. There is one big difference between British bikes and Chinese bikes though.

G: Oh, what is it?

N: We have lights on them. It's actually illegal to ride a bike at night without lights.

G: We would never use lights. Perhaps our night vision is better than yours!

习惯用语 1

① mountain bike: 山地自行车

N: 路上的自行车可真多!

G: 我们过去不是以"自行车王国"而著称嘛。不过现在路上的自行车不像原来那么多了。

N: 在英国,我们可没有这么多自行车,自行车道也很少。

G: 为什么?

N: 虽然大多数小孩子会有自行车,但等他们到了能开车的年龄,他们就会去买汽车了。

G: 英国的学生有自行车吗?

N: 要是有的话,也很可能是山地车。但大多数学生都有汽车。

G: 那他们肯定特有钱!

N: 不是的。他们会买一辆便宜的二手车。

G: 我听说牛津和剑桥的学生就骑自行车,是这样吗?

N: 是的,尽管牛津和剑桥都是很小的城市,且学校就坐落在市中心。但骑车更多的是出于一种传统而非其它原因。英国自行车与中国自行车之间也有一个很大的不同点。

G: 哦? 是什么?

N: 我们的自行车上有车灯。夜间骑没有灯的自行车是违法行为。

G: 我们从没用过车灯。也许是我们这里晚上的视野比你们的要好一点吧!

Questions 1

1. Do you have a mountain bike?
2. Do you have lights for your bike? Why? Why not?

Dialogue 2

Nick: The campus I live on is really small but it would be quite beautiful if it weren't for one thing.

Gloria: Oh what's that?

N: Bikes. It's really too small a campus for bikes yet you see them everywhere. And many of them seem to have been abandoned. Abandoned bikes are an **eyesore**[①] and turn what should be a pleasant sight for the eyes into a junkyard.

G: I've just been reading in the paper how several truckloads of abandoned bikes were cleared from the Peking University campus and most of these bikes used to belong to students.

N: I used to teach there and Peking University is the most beautiful campus in Beijing and many tourists like to visit it at the weekends. But what will they think of Beida and its students when they see rusting bikes carelessly left to rot everywhere.

G: Good job the university authorities cleaned it up. Do you think the same thing should happen here?

N: I think I would be a bit more radical than that and have them banned altogether.

G: That sounds drastic! What's your reasons?

N: Well, because it's a small campus students don't need bikes. Secondly, students wouldn't complain about their bikes being stolen because there would be no bikes to steal!

G: Do you really think the university and the students would agree to a ban?

N: I think the first thing that should be done would be to get rid of the abandoned bikes. Then the university should carry out a survey to see how much the students used their bikes and also find out how many bikes are stolen every year.

G: Sounds sensible to me.

习惯用语 2

① eyesore: 看上去不顺眼的东西

N: 我住的那个校园真的很小，不过要不是因为一样东西，它还是很美丽的。

G: 哦，是什么？

N: 自行车。校园面积太小了，四处都是自行车。而且很多车似乎都是被人遗弃的。这些遗弃的自行车真是大煞风景，把本来美丽的校园变得像垃圾场一样。

G: 我刚刚看报上说，从北大校园里清理出了好几卡车的废弃自行车，大多数都是属于学生的。

N: 我在那里教过书，北大是北京最美的校园，许多游客都喜欢在周末的时候到那里逛一逛。可当他们看到了这些四处丢弃、无人问津的生了锈的自行车，对北大和北大学生会怎么想呢？

G: 好在学校把它们都清理了。你觉得类似的事情在我们学校会发生吗？

N: 要是我的话，就再坚决一点儿，从此在校园里禁止自行车。

G: 听起来很极端呀！你的理由是什么？

N: 因为校园面积很小，根本就不需要自行车。这样一来，学生也就不用再抱怨自行车被盗了，因为这儿没有自行车可偷。

G: 你觉得校方和学生们会认同这种禁令吗？

N: 我认为首先应该清理掉那些废弃的自行车。然后要做一个调查，看看有多少学生使用自行车，每年被盗的自行车数量又是多少。

G: 嗯，这听起来挺有道理的。

Questions 2

1. What changes would you make to your campus?
2. Would you ban all bikes from your campus? Why? Why not?

Dialogue 3

Nick: I really like the idea of bicycle lanes here. It's just a pity it doesn't work out in practice.

Gloria: Why do you say that?

N: Not only do cyclists ride the wrong way but they also do not stick to bicycle lanes. They use roads, pavements and any ground where two wheels can go.

G: We are very inventive when it comes to getting to where we

Travel 旅行

want to go!

N: I suppose that's one way to look at it. Can you tell me why when the traffic lights are at red and all the cars stop the bikes don't. I've nearly been **run over**① by cyclists at zebra crossings because of that.

G: Everybody does that. I don't know why they don't stop.

N: Another thing is that Chinese cyclists seem to have no spatial awareness, no concept of what is in front, to the back or sides of them. I have often seen riders cycling along not looking where they were going, oblivious of their surroundings. Coming from a minor road to a major one they shoot out at full speed, irrespective of oncoming traffic. Chinese cyclists seem to believe they are invincible, instantly noticeable and immune to any danger. Until they have an accident of course!

G: Considering the huge number of cyclists we have I think that we have very few accidents so cyclists must have an angel looking over them!

N: That must be so but I'm still not buying a bike!

习惯用语3

① run over: 撞倒

N: 设置自行车专用道这个主意真的很不错，只可惜这实际上并没有什么作用。

G: 怎么这么说？

N: 骑车人不但骑车乱行，还不进入自行车专用道。马路、人行道，只要两个轮子能骑到的地方，就有他们的身影。

G: 一提到出行，我们就很有创意。

N: 这只是一方面。你能不能告诉我，为什么红灯亮了，汽车都停下来了，自行车却不停？我曾经就差点儿在斑马线上被骑车人撞倒。

G: 每个人都这样做。我也不知道他们为什么不停车。

N: 还有一点，中国的骑车人似乎都不太有空间感，分不清前后左右。我经常会看到有人骑着车，不是往前看，而是环顾四周。当从辅路转向主路的时候，也不知道减速，完全不顾迎面开来的车辆。他们似乎觉得自己是无敌的、能够很快就被发现，永远不会发生危险。直到他们真的出事故了才行。

G: 我们骑车的人那么多，事故率却很低，肯定有个守护天使在守护着他们。

N: 确实是这样，但我至今也没买自行车。

Questions 3

1. Do you stop at traffic lights? Why? Why not?
2. How many accidents have you had while riding your bike?

Travel 旅行

20   行人

Pedestrians

|| Background Information ||

11% of Beijing pedestrians crossed the streets without using zebra crossings in 2006.

A recent survey on the time it took 35 men and women to walk along a 60 ft stretch of pavement found that people in Singapore walked the fastest, covering the ground in 10.55 seconds. In China, people in Guangzhou came fourth (10.94) and Taipei came twenty-third (14.00). Men are generally 25 percent quicker on their feet than women.

2006 年，北京有 11% 的路上行人在过马路时不走斑马线。

最近针对 35 名男士和女士在走过长 60 英尺的路程所花费的时间展开了一项调查，调查结果显示，新加坡人走得最快，走完这段路程花费 10.55 秒。中国广州的人位列第四，花费 10.94 秒。台北排在第 23 位，花费 14 秒。男士一般比女士步行速度快 25%。

背景信息

Dialogue 1

Nick: Why do they have these metal fences running down the central division of roads? Is it to stop cars from **oncoming traffic**[①]?

Rachel: No, it's actually to stop pedestrians from crossing the road.

N: I've noticed that pedestrian crossings here are different from ones in the U. K.

R: In what way?

N: Well when the lights are green then pedestrians have **the right of way**[②] but here you often find traffic that is turning right continue to do so.

R: I'm afraid that here in Beijing cars are the kings of the road and they behave as if they own it.

N: Makes life difficult for us pedestrians then. I have to say I often feel a little worried if I have to cross a busy road with no subway or footbridge. I just try to keep inside a group of people, at least you can get a little protection that way.

R: We've learned to cross the road whenever and wherever we can. There are few subways here where you can cross under the road.

N: And often the pavements are too full of people, so I've seen many just walking along the road.

R: Yes, it's acceptable for us to walk on the road. There are so many people in China, we just walk wherever we can. I guess I've been doing it for so long, I'm used to it.

N: Have you ever crossed a road illegally?

R: I used to, yes, just to save time. But then, last summer, I saw an awful accident where a man was hit by a Jeep. He died instantly and there was blood all over the road.

N: Well, that would certainly put me off for life!

Travel 旅行

① oncoming traffic: 迎面过来的车辆
② the right of way: 路权，过马路的时候汽车必须停下来

N: 为什么要在马路中间设置铁栅栏？是用来阻止那些对面驶来的车辆吗？

R: 不是，其实是为了防止行人横穿马路。

N: 我注意到了，这里的行人过马路与英国不太一样。

R: 怎么不一样了？

N: 当绿灯亮起的时候，行人拥有路权，可以前行。但在这儿右转的机动车却还在行驶。

R: 恐怕在北京，汽车就是马路上的老大，肆无忌惮，似乎马路是归他所有的。

N: 这可让我们行人难办了。我很害怕穿过那种既没有地下通道也没有过街天桥的马路。我会使劲地钻进人群，那样才觉得有点儿保护。

R: 我们已经学会如何不受时间地点限制地过马路了。有些地方有地下通道，你也可以从地下穿过马路。

N: 很多时候人行道都走不下人了，我看到很多人索性就沿着马路走。

R: 是的，沿着马路走还是可以接受的。中国人这么多，能走的地方我们就走。我也经常这么做，我已经习惯了。

N: 你有没有违法穿越过马路？

R: 过去有过，为了省时间。但去年夏天，我看到了一起严重的事故，一个男人被一辆吉普车撞倒在地，当场死亡，马路上全是血。

N: 哦，为了生命安全，我是不会那么做的。

Questions 1

1. Do you ever cross the road illegally?
2. If you were the mayor of your city how would you deal with this problem?

Dialogue 2

Nick: On campus here I've noticed that if there are pavements then often there are trees in the middle of them or bikes are parked on them and so pedestrians have to walk in the road.

Rachel: Yes, that's very common here. We are all trying to make Beijing a greener place before the Olympics, so we are trying to plant more trees and preserve the ones we already have.

N: Actually, universities seem to be some of the greenest places in Beijing, but couldn't they have planted the trees in other places, like in a park area so people wouldn't have to walk on the road to go around them?

R: I have never really thought about it. But I do agree with you about the bikes, though.

N: There seems to be bikes everywhere! How come there are so many, especially here on campus?

R: Well, we students are usually quite poor, so we can't afford any other type of transport. None of my classmates own cars, most of them have no idea how to drive.

N: When I was at university in England, many of my classmates had cars so they could drive to and from classes. They needed cars because they didn't live on campus, so the pavements were clear, but the car parks were full.

R: So I guess the pavements here are very different, but you get used to it. You must learn to be nimble so you can squeeze through small spaces without knocking over a bike.

N: Yes, I can see that. My goodness, if you knocked down one it would be like **a domino effect**[①] and the whole lot will go!

R: Haha! So maybe you should always walk on the road, you are so big and clumsy I think maybe you could beat any oncoming traffic anyway. There's no need for you to be scared of cars!

N: Still, I feel it's strange I have to walk on the road, though. Pavements are meant to be for people, not for bikes.

Travel 旅行

R: So if I were you, I'd take up Tai Qi so you can improve your balance, and that way no accident will happen.

习惯用语 2

① a domino effect: 多米诺效应

N: 我发现在这儿的校园里，人行道上通常都会种树，有人还把自行车停在人行道上，这样一来，行人就只能沿着马路走了。

R: 没错，这很常见。为了迎接奥运，我们要让北京变成一个绿色都市。所以，在保护原有树木的基础上，我们要尽可能地多种树。

N: 其实，大学基本上是北京最绿的地方了。他们就不能把树种到别的地方吗？比如公园。省的人们还得为了绕开树而到汽车道上走。

R: 我还真没想过这个问题。不过我倒是很同意你对自行车的看法。

N: 到处都是自行车，怎么会有这么多呢？尤其是在学校里。

R: 学生都很穷啊，我们负担不起其他形式的交通工具。我的同学都没有汽车，好多人连开车都不会。

N: 我在英国上大学的时候，很多同学都有车，可以开车上下学。他们需要开车，是因为他们不住在学校里。所以英国大学的人行道上很干净，不过停车场却爆满。

R: 我想这里的人行道一定很特别，但你已经习惯了。你得动作敏捷一点儿，挤过狭小的空间而不碰到那些自行车。

N: 是的。天哪，要是你碰倒了一辆，那就会是多米诺效应，整个一排都会倒下去！

R: 哈哈！也许你就应该一直在马路上走。你个头儿这么大，动作又慢，说不定会撞倒迎面开过来的车，所以你不必害怕汽车！

N: 我还是觉得很不能理解我为什么得在机动车道上行走。人行道就是为行人准备的，而不是为自行车。

R: 如果我是你，我就去练太极以提高平衡能力，那样就不会发生危险了。

Questions 2

1. Do you park your bike in a considerate way?
2. Do you usually walk on the pavement or in the road?

Dialogue 3

Michelle: What's wrong with you today? You've been grumpy ever since we came out of the subway.

Becky: I'm sick and tired of[①] trying to avoid the cars in Beijing, especially when I'm shopping. I don't like going to just one shop, I want to walk around and browse. I wish I could go somewhere where there are no cars allowed. It seems that people aren't as important as cars here.

M: Actually, there are two great places I know of where cars aren't allowed and you can shop **until your heart's content**[②]! One place is Wangfujing, probably the most famous shopping street in Beijing.

B: I've heard, of Wangfujing, but I didn't know cars weren't allowed. I just thought that it would be like the other shopping areas where you **take your life in your hands**[③] every time you leave a shop.

M: Also, Qianmen, near to Tian'anmen Square, has recently become a **pedestrianised zone**[④]. I think maybe Qianmen will be more interesting because even though it's newer, it's not as modern.

B: You mean they haven't just demolished everything? That's another thing that makes people have to walk in the path of cars, there are building sites everywhere.

M: Yes, they have tried to preserve at least a hundred of the old shops. So, that way, we can get a more traditional atmosphere there than at Wangfujing.

B: I think we should try Wangfujing first, it's more famous. It will be great to go somewhere with no cars using me as target practice!

M: Well, some special cars are allowed. They are small, open buses for shoppers who don't want to walk. But they aren't free, you must pay and to be honest the drivers can be a little crazy.

B: They have crazy drivers! What's the point in that? If they say the road is only for pedestrians, it should only be for pedestrians. Who uses these little buses, anyway, locals?

M: Actually, mostly tourists. Not just foreigners, either, but tourists from all over China. To be honest, I'm a Beijinger, but I've hardly been there. Locals tend to shop in other places that aren't pedestrianised, we are used to all the cars and the crazy taxi drivers.

B: I'd certainly like to go and take a look, but I don't think I'll bother with the shoppers' bus. It sounds like I'm still going to need **eyes in the back of my head**⑤ if I'm going to shop there safely.

习惯用语 3

① I'm sick and tired of: 我很讨厌⋯
② until sb.'s heart's content: 尽情地
③ take sb.'s life in sb.'s hands: 冒着受伤的风险
④ pedestrianised zone: 商业步行区
⑤ eyes in the back of sb.'s head: 脑后长眼（可以看到360度的景象）

M: 你今天是怎么了？从一出地铁，你的脸色就不好。

B: 在北京，我真的很讨厌躲让汽车，尤其是在我购物的时候。我不喜欢只逛一家店，我想多转转，多看看。真想去个没有车的地方。在这儿，似乎觉得人不如车重要。

M: 其实我知道有两个禁止汽车通行的地方，你可以在那里尽情地逛。其中之一就是王府井，这大概是北京最著名的购物街了。

B: 我听说过王府井，但我还真不知道那里禁止汽车通行。我还以为那里跟其它购物区一样，你一走出商店就得冒生命危险呢。

M: 还有前门，离天安门广场不远，那里刚刚变成商业步行区。我想前门可能会更有意思，尽管是刚刚建立，不过不那么现代化。

B: 你是说还没有遭到破坏？四处都是建筑物，这也是把行人推向汽车道的原因之一。

M: 是的，那里尽力保留下来了至少一百个老商铺，所以传统的气息比王府井要浓郁。

B: 我觉得我们还是应该先去王府井，那里比较有名。想去哪儿就去哪儿，真是太好了，省得成了汽车的靶子。

M: 也有一些专用车是允许在其中穿行的，就是那些专门为不想走路的购物者准备的小型的、开放空间的公共汽车。但那并不是免费的，而且说实话，司机开得也有点儿疯狂。

B: 疯狂的司机？为什么啊？既然是步行街，就应该只允许行人走。谁使用这些小公共啊，是当地人吗？

M: 其实大多是游客在使用。中国各地的游客以及老外都用。说实话，我是个北京人，但却很少去那种地方。北京人反而不愿意去步行街购物，我们已经习惯了四处乱开的汽车和疯狂的出租司机。

B: 我很想去逛一逛，但我却讨厌那些购物专车。感觉我还是得能够眼观六路，才能保障购物安全。

Questions 3

1. What do you think of pedestrianized zones?
2. Are taxi drivers really crazy?

T r a v e l 旅行

21 外貌
Appearance

‖ **Background Information** ‖

According to a 2006 survey by the Ministry of Education the average height of male students in 2005 was 172.5 cm. The average height of female students was 160.5 cm.

The average height of U. K. males is 175.5 cm and for females it is 162 cm. British men average 79.75 kilos in weight and 66.7 for women. The average British woman's figure is 100.78 cm, 84.06 cm and 87.37 cm. The average British male's foot is 26.68 cm long.

2006 年，教育部的一项调查显示，2005 年男生的平均身高为 172.5 厘米，女生的平均身高为 160.5 厘米。

同期英国的男女平均身高分别为 175.5 厘米和 162 厘米。英国的男女平均体重分别为 79.75 公斤和 66.7 公斤。英国女士的平均三围分别为 100.78 厘米，84.06 厘米和 87.37 厘米。英国男人的脚平均长度为 26.68 厘米。

背景信息

Gail: What beautiful blue eyes you have!

Nick: Thank you and you have beautiful brown eyes too.

G: Brown eyes! No they're black!

N: Have you got a mirror?

G: Yes, I've got one in my purse.

N: Look at your eyes in it. You'll see that the pupil—that's the centre of your eyes—is black while the iris is brown. My irises are blue so I have blue eyes. Your irises are brown so you have brown eyes.

G: Well, I was always brought up to think I had black eyes!

N: In English, if you say you have a black eye it means that someone's hit you in the face causing your eye to be swollen and discoloured.

G: I see.

N: Another way you misuse the word black is when someone has been out in the sun for a long time and you say that they are black.

G: What's wrong with that?

N: In England we would say that they are brown or tanned. You should do the same too.

G: OK in future I'll try not to use the word black so often!

G: 你的蓝眼睛可真好看!

N: 谢谢,你棕色的眼睛也很迷人。

G: 棕色? 不对,是黑色的!

N: 你有镜子没有?

G: 我钱包里有一个。

N: 看你自己的眼睛,中间的瞳孔是黑色的,周围虹膜是棕色的。我的虹膜是蓝色的,所以我有一双蓝眼睛。而你的是棕色的,所以你是棕色的眼睛。

People 人

G: 我一直以为我的眼睛是黑色的！

N: 在英国，要是你跟别人说你有一双黑眼睛，人家会以为你被击中了面部，导致你的眼睛肿胀，变色了。

G: 我明白了。

N: 对"black"这个词，你还有一个误用之处。如果一个人在太阳下面晒了很久，你会说他们很"black"。

G: 这有什么错吗？

N: 在英国，我们用"brown"或者"tanned"来形容皮肤被晒黑了。你也应该这样说。

G: 好的，那我以后还是尽量少用"black"这个词吧。

Questions 1

1. What colour are your eyes?
2. Do you prefer a pale or darker skin? Why?

Dialogue 2

Rose: What's one of the biggest differences you've noticed between life in China and life in England?

Nick: During a hot day in summer everyone **covers up**① and out comes the umbrella! In England we only use umbrellas when it's raining!

R: We don't like to be black. That's why we protect ourselves from the sun. We think that a pale skin is more beautiful than a dark one.

N: Again that's a big difference between us. We think a tan makes us look more healthy and attractive. It's why we like to wear as little as possible during the summer and why we go to hot countries or beaches on holiday.

R: They say that beauty is in the eye of the beholder! So we have different views on what beauty is.

N: Yes, especially when it comes to pretty girls! Tell me what you think a pretty girl looks like?

R: Well, she should have large eyes with double eyelids, a high nose with straight sides, a V shaped face and a small mouth like a cherry.

N: We Westerners think that a pretty Chinese girl is one who has small eyes with single eyelids, high cheekbones and a big mouth.

R: A big mouth! Wah so ugly!

N: We call it **bee stung lips**② and think it's very sexy.

R: Like Angelina Jolie, right?

N: Right! So to me you are not a very attractive girl!

R: What a pity!

习惯用语 2

① cover up: 穿很多衣服
② bee stung lip: 蜂蜇唇，嘴唇饱满

R: 在中国生活与在英国相比，你觉得最大的不同是什么？

N: 在中国，一到夏天人们就捂得严严实实的，出门还打伞。在英国，我们只有下雨的时候才会打伞。

R: 因为我们不想被晒黑啊，所以才要防晒，我们觉得嫩白的皮肤比黑色的好看。

N: 这又是我们之间一个很大的不同。我们觉得棕褐色的皮肤看起来更健康，更迷人。所以到了夏天，我们会穿得尽可能少，假期还要到热带国家或者海边去晒太阳。

R: 人们常说，好看与否是由旁人的眼光决定的。看来我们的审美观点有所不同。

N: 对，尤其是对漂亮女孩的看法。你认为什么样的女孩才算好看？

R: 嗯，大眼睛、双眼皮、高鼻梁、瓜子脸，还有樱桃小口。

N: 我们西方人认为，漂亮的中国女孩应该是小眼睛、单眼皮、高颧骨、大嘴巴。

People 人

R: 大嘴巴？那多丑啊！

N: 我们管那叫做"蜂蜇唇"，觉得那样很性感。

R: 像安吉丽娜·朱莉那样，对吧？

N: 没错！所以，对我来说，你不是一个有吸引力的女孩。

R: 真遗憾啊！

Questions 2

1. What's the biggest difference you've noticed between Westerners and Chinese?

2. What do you think a pretty Chinese girl looks like? A pretty Western girl?

Dialogue 3

Nick: Do you know why Chinese people are normally smaller than Western people?

Rose: Well, I think that's because of their genes.

N: No, I don't think that's the reason.

R: Then what's the reason do you think?

N: I think it's due to better food and nutrition. Food is very plentiful everywhere and there's much more variety.

R: No, I don't think so. We believe whether a person is fat or thin, tall or short is due to his mother.

N: I think you may be right but eating junk food like McDonald's and KFC does not help. In the West we eat too much of the wrong foods with the result that at least 25 percent of British people are obese.

R: Chinese food is a lot healthier so that's why you don't see a lot of fat Chinese people.

N: I've certainly noticed that since I eat Chinese food every day that my weight has dropped.

R: But you haven't got any smaller, though!

N: Well, my mother is small but my father is tall so maybe your theory only applies to Chinese!

R: I wonder what your children will be like if you marry a Chinese girl!

N: They'll probably have one brown eye and one blue eye!

N: 你知道为什么大多数中国人都比西方人矮小吗?

R: 我想是由于基因决定的吧。

N: 不，我觉得不是。

R: 那你觉得是什么原因?

N: 我想是因为西方人的食物和营养比较好。食物充足，选择的余地也很大。

R: 我可不这么认为。我们相信，一个人的高矮胖瘦跟他的母亲有很大关系。

N: 也许你是对的，不过吃麦当劳、肯德基那种垃圾食品没什么用。在西方，我们吃了好多不恰当的食物，以至于 25% 的英国人都过度肥胖。

R: 中餐就比较健康，所以你很难看到那么多中国胖子。

N: 我也注意到了，自从我开始每天都吃中餐以来，我的体重确实下降了。

R: 可你的个头儿可没有变小啊!

N: 我妈妈个子矮，不过我爸爸很高。也许你的理论只适用于中国人。

R: 要是你娶一个中国女孩，我真想知道你们的孩子将来会比较像谁。

N: 说不定他们会有一只蓝眼睛，一只棕眼睛。

Questions 3

1. Is size dependent on genes or nutrition?
2. What do children of East-West marriages look like?

People 人

22 礼节

Manners

||Background Information||

Beginning in February 2007 every 11th day of the month is to be known as "Queuing Day". The reasons why people don't queue are because

- others don't 57%
- I'm in a rush 35%
- I never knew I had to 8%

从 2007 年 2 月开始，每月的 11 日都被指定为"排队日"。人们之所以不排队，原因是：
- 别人不排队 57%
- 我有急事 35%
- 从不知道需要排队 8%

背景信息

Dialogue 1

Nick: I really hate following people into shops.

Carole: Why?

N: You know that during the summer there are often these plastic strips that hang down and you have to push them aside to enter.

C: Yeah, so what?

N: Well most people never seem to think of people behind them so often I get hit in the face by them.

C: Perhaps they never noticed you.

N: In England we are taught to open doors for other people so I always do that.

C: We are taught good manners too but I suppose we mainly do that for people we know.

N: Another thing I hate are lifts because people never seem to wait for people to exit before they enter. It always surprises me that they seem to automatically think the arriving lift is going to be empty and so they stand before the doors so that when they open they are ready to walk in.

C: Maybe they are in a hurry.

N: Maybe but sometimes I point it out to them and they say "Sorry, sorry."

C: Do British people say "Sorry" a lot?

N: All the time! Even if it's not our fault.

N: 我特别讨厌跟在别人后面进商场。

C: 为什么?

N: 你知道,到了夏天,商场的门口都挂有那种塑料的条形门帘,要想走进商场,就得把这些帘子弄到一边去。

C: 对啊,那怎么了?

N: 很多人都不会考虑他们身后的人,所以我经常被帘子打到脸。

C: 可能是他们没有注意到你。

N: 在英国，我们受教育要给别人开门，我经常这样做。

C: 我们也被教导要有礼貌，但我估计大多数人是对自己认识的人有礼貌。

N: 还有就是我特别讨厌电梯。那些等电梯的人都不等里面的人出来就往里走。让我感到吃惊的是，他们似乎总觉得电梯是空的，老早就站在电梯门前，门一开就准备冲进去。

C: 也许他们有什么急事儿。

N: 也许吧，不过有时我向他们指出来，他们就说对不起。

C: 英国人经常说对不起吗？

N: 说啊，即便不是我们的错，我们也会说对不起。

Questions 1

1. What's your behaviour like when you enter a building through a door?
2. What do you do when you enter or leave a lift?

Dialogue 2

Nick: Why is everyone queuing at the bus stop? They don't normally do that.

Carole: Today's the 11th. Every month when it's the 11th we have to queue otherwise we will get into trouble.

N: Why the 11th?

C: Because the number 11 looks like two orderly lines.

N: I think it's a bad idea just to do one day a month because people won't do it the other thirty days.

C: I agree. Do people queue in the U.K.?

N: All the time. Ever since bus stops came into use in the 1930's we queue so it's become an **ingrained habit**[1] now.

C: We've started to queue in banks and post offices.

N: About time! I hope you stop spitting, too!

C: I know, it's embarrassing to be Chinese when someone does that.

N: I hate the hacking noise that precedes it when someone clears their throat.

C: You know that Beijing is the most polluted city in the world so people can't help getting dust in the throats and noses.

N: True, but I just wish they would do it quietly!

习惯用语 2

① ingrained habit: 根深蒂固的习惯

N: 大家干嘛都在车站排队啊? 平时可不是这样。

C: 今天是 11 号啊, 每个月的 11 号我们都得排队, 不然就会有麻烦。

N: 为什么非得是 11 号?

C: 因为 11 这个数字看起来就像是两条有秩序的队伍啊。

N: 一个月就这么一天排队日, 可真不是什么好主意。每个月剩下的日子人们就不会排队了。

C: 我也这样觉得。在英国, 人们排队吗?

N: 干什么都排队。自从 20 世纪 30 年代公共汽车开始投入使用的时候, 我们就开始排队, 现在已经是一种习惯了。

C: 我们在银行和邮局现在已经开始排队了。

N: 早就该这样嘛。还有, 我希望你不要随地吐痰!

C: 我知道, 看到有人随地吐痰, 作为中国人我都感到很尴尬。

N: 我真的很讨厌别人清嗓子时发出的声音。

C: 北京是世界上污染最严重的城市, 所以人们的嗓子和鼻腔里难免会有脏东西。

N: 没错, 我只是希望他们能小声一点儿。

Questions 2

1. Do you always queue?

2. What do you think of spitting?

People 人

Dialogue 3

Carole: I hear the government has issued a good manners leaflet for all Chinese tourists going abroad.

Nick: Yeah, it seems there's been lots of complaints about their behaviour. I heard they were particularly bad at Hong Kong's Disneyland with their spitting, smoking, loud talking and sitting down everywhere.

C: There were even etiquette police out in force during May Day and over fifty tourists were fined for spitting or littering in Tian'anmen Square.

N: I gather the government is worried about people giving visitors to the Olympics a bad name. I often see men rolling up their trouser legs or shirts and showing a lot of skin.

C: It's only men who do that. We ladies prefer to cover up during hot weather.

N: Many countries are now banning smoking in public areas. I know Ireland and England has.

C: Many Chinese men smoke and some public places now ban smoking but to be honest many people just ignore the signs.

N: Learning good manners takes time and really it's something parents should teach rather than governments enforce.

C: What about British tourists?

N: Actually, we're just as bad. We are known for our drunken behaviour in places like Spain.

C: So if British and Chinese tourists met together then we would both be the **tourists from hell**[①]!

N: But if we met and borrowed the good points from each other and eliminated the bad then we would be tourists from heaven!

① tourists from hell: 最糟糕、最差劲的游客

C: 我听说政府为所有出境旅游的中国人发了一本文明行为宣传册。

N: 是的，有些游客的行为确实不怎么样。我听说他们在香港迪斯尼乐园的表现尤为糟糕，随地吐痰、抽烟、大声讲话，还到处乱坐。

C: 在五一期间，甚至还有"文明警察"巡逻执勤。在天安门广场，有 50 多名游客就因为随地吐痰、乱扔垃圾而被罚了款。

N: 我想政府可能是怕人们给奥运时来中国的外国游客留下不好的印象。我经常可以看到男人们挽起裤腿或上衣，身子露在外面。

C: 只有男人才那么做。我们女的倒是更愿意在夏天捂得严实一点儿。

N: 很多国家都禁止在公共场所吸烟，爱尔兰和英国就是这样。

C: 很多中国男人都吸烟，现在一些公共场所也禁止吸烟，但是说实话，很多人对这种禁令根本就视而不见。

N: 养成良好的行为举止是需要时间的。这是父母教育的一种责任，而不应该是政府的强迫命令。

C: 英国的游客怎么样？

N: 其实我们也差不多，我们以酒后滋事而闻名，就比如在西班牙。

C: 所以，要是让中英两国的游客聚到一起，那就是来自地狱的魔鬼！

N: 但如果我们彼此取长补短，相互借鉴对方的优点，改正缺点，我们也会成为来自天堂的天使。

Questions 3

1. Which bad manners do you hate the most?
2. How would you get people to be better mannered?

People 人

23   服 装

Clothes

‖ **Background Information** ‖

Clothing styles have changed tremendously so much over the last fifty years that by looking at a picture of a person you can usually tell which decade it was taken in. However, one common constant seems to be denim jeans. You can find these in many different styles from figure hugging to baggy low hung.

在过去的 50 年间，服装的款式发生了巨大的变化，以至于你只要看一看照片上的人，基本上就能说出这照片大概是什么时候照的。但有一样东西一直流行至今，它就是粗斜纹牛仔裤，它有很多不同的款式，比如有紧身的，也有宽松低腰的。

背景信息

Dialogue 1

Nick: Why do Chinese students when they graduate wear Western-style gowns?

Mandy: I think this started in the 1920's as a universal symbol indicating academic achievement and gradually spread to all of China's universities.

N: I think it looks odd for Chinese to wear since it originates from the clothes that western monks wore in the Middle Ages. Why don't you change to a more Chinese style?

M: Actually I think we are. I know that Peking University is considering changing to Chinese dress and has asked students to submit designs.

N: What do you think they will look like?

M: The favourite seems to be the Hanfu which was the pre-17th century traditional dress of the Han majority Chinese group.

N: What does it look like?

M: It is a gown with wide sleeves, layered robes and crossed collar-bands.

N: I hope that design catches on.

M: According to a web survey of over 30,000 people 80 percent like it so it is now up to the universities to implement it.

N: I'm looking forward to seeing you graduate in it.

M: I think that will be the proudest moment of my life. But if students wear Chinese dress then professors should too.

N: You're right but I don't know if it will happen in my lifetime!

N: 为什么中国学生在毕业的时候要穿那种西式的学位袍呢？

M: 我想这得从 20 世纪 20 年代说起，那时候，这种服装普遍作为一种具有学术成就的象征，慢慢地，就传遍中国所有的大学了。

N: 我觉得中国人穿这种衣服感觉怪怪的。因为这种衣服起源于中世纪西方传教士的服装。你们干嘛不穿那种更加中式的衣服啊？

M: 其实我也这么觉得。我知道北大就正在考虑让学生在毕业典礼上穿中式服装，都号召学生们递交服装设计稿了。

N: 你觉得那会是什么样？

M: 最受欢迎的似乎是汉服，就是 17 世纪早期的那种传统汉族服饰。

N: 是什么样的？

M: 是一种袖子宽大的分层礼袍，胸前有一个交叉的领子。

N: 我希望他们的设计能够时尚些。

M: 在网上，有一个 3 万多人参加的网络调查，80% 的人都喜欢这种汉服，所以现在就等着校方将其付诸实践了。

N: 那我就等着看你毕业的时候穿汉服的样子了。

M: 我想那会是我人生中最自豪的时刻。不过，要是学生们穿了中式服装，那教授们也应该这样穿。

N: 你说的没错，不过，不知道我能不能活到那一天啊！

Questions 1

1. When you graduate would you prefer to wear Western-style or Han-style gowns? Why?
2. Do you think that Chinese professors should switch over to Chinese academic dress?

Dialogue 2

Tori: Here it is! I've been looking for this kind of short leather coat for ages.

Rick: Oh, really? But this one is too old, not only does it look like it's been worn by lots of generations, but it's also like 70's **rocker's gear**①.

T: You don't know! I love this style, it's vintage, and toooo cool. You can't get it from any normal shops. Some specialists sell this kind of stuff, but you get lots cheaper in a charity shop.

R: I know this kind of shop, run by some charity groups or foundations. When I lived in Edinburgh, one of my friends worked in a charity shop. She also bought some clothes and jewellery from the shop.

T: Have you got anything?

R: Not really. Actually, I've never been in any second-hand shops. In China, apart from cars and houses, I can't think of anything that people will buy that's second-hand. In our culture, only if you're too poor to afford new stuff, you may buy used things.

T: You mean you are loaded.

R: No. I mean we have a different view about old things.

T: OK, we say that someone's junk is another's treasure. All of the stuff in a charity shop is donated by local residents and other shops. Most of them are in good condition; actually they are not necessarily very old, some things are even new. Some people just changed their minds quicker than others. Staff here will classify the donations, clean them when needed and put reasonable prices on. It helps people like me to get decent goods with decent prices. Also, all the money the shop makes will go to the organisation to help others who need help.

R: Really? That's great. I can save money at the same time as helping people. OK, should I start with new stuff? I want some trousers. But not vintage though.

T: Well, go to the back of the shop, you'll find some brand new trousers, they're all in weird sizes, but never mind, you'll get some for your short fat legs.

R: You'd better go with me, and you may find some matching "vintage" for your giant feet!

T: Oh, you have learnt a lot today!

People 人

T: 终于找到了！我找这种短款的皮外衣已经好久了！

R: 真的？这也太老了，好像很早以前就有人穿了，看起来有点儿像 70 年代摇滚歌手穿的。

T: 你不懂！我就喜欢这个款式，复古，而且很酷！在一般的店里都买不到。有专门卖这种衣服的店，不过福利商店里卖得更便宜。

R: 我知道这种商店，都是由慈善组织或者基金会开办的，我在爱丁堡居住的时候，有一个朋友就在福利店里工作，她也会从店里买些服装、首饰什么的。

T: 你买了吗？

R: 没有。其实我从没去过二手商店。在中国，除了房子和汽车，我想不出什么东西人们会去买二手货。在我们的文化里，除非你很穷，买不起新东西，才会去买二手的。

T: 那你意思就是你很殷实富足喽？

R: 那也不是，我的意思是，对于旧物件，我们有不同的看法。

T: 我们觉得，有些人的垃圾，对其他人来说可能就是宝贝。福利商店里的东西都是当地居民和其他商店捐赠的。大多数东西的状况都很不错的，不一定都很旧，有些东西甚至很新。只不过有些人的主意转变得比较快罢了。这里的员工会把捐赠品进行分类、清理，然后贴上合理的价钱。这就可以让我这种人用好价钱买到好东西。同时，商店的收入都会上交组织，去帮助其他需要帮助的人。

R: 真的？太好了！在帮助他人的同时还可以为自己省钱。我可以在这里买些新东西吗？我想挑几条裤子。可不要这种古董货啊。

T: 就在商店的后面，你会找到很多全新的裤子，尺码都很夸张。不过别着急，肯定能找到适合你的小短腿儿的。

R: 跟我一起过去吧，给你的大脚找双合适的"古董"。

T: 哦，现学现卖啊你！

Questions 2

1. Do you ever buy second-hand clothes? Why? Why not?
2. What do you do with your old clothes?

Dialogue 3

Nick: It's funny[①] that the vast majority of Chinese students wear Western-style clothing.

Mandy: Why funny? It's perfectly normal.

N: I suppose so. It's just that I really like Chinese clothes. Especially the clothes that some of my female students wear. They look really fashionable.

M: But most students think Western clothes are more fashionable, especially designer clothing.

N: In the West most of our clothes have labels saying Made in China. But they are all Western designed. I think it is time China started to make its own clothes fashionable with labels saying Designed in China.

M: That sounds like a great idea! If I were a fashion designer I would start on that right away!

N: You could be an entrepreneur and hire people to do that.

M: Do you ever wear Chinese clothes?

N: I have a red silk jacket but to be honest I really look stupid in it. A bit like Da Shan looks in his!

M: Ha ha. I know what you mean!

N: Some Chinese clothes do not suit us but I like Chinese shirts. Now they would really sell well.

M: Maybe I should open a shop.

N: And online too, don't forget!

People 人

① It's funny: 真奇怪

N: 几乎所有的中国学生都穿西式服装，真奇怪。

M: 有什么奇怪的？很平常啊。

N: 我真的很喜欢中国的服饰，尤其是我的一些女学生们穿的，看起来真的很时髦。

M: 但大多数学生还是觉得西式服装更时尚，尤其是那种专门由服装设计师设计的衣服。

N: 在西方，我们很多服装的标签上都写着"中国制造"，但都是西式的设计。现在，中国应该设计自己的时尚服装，然后在标签上注明"中国设计"。

M: 这主意听起来不错啊！我要是个服装设计师，肯定马上就干。

N: 你可以当个企业家啊，雇人来做这些工作。

M: 你穿过中式服装吗？

N: 我有一件红色的丝制夹克，不过说实话，我穿起来很傻。就好像大山穿他的那件红夹克一样。

M: 哈哈！我明白你的意思了！

N: 有些中式服装不太适合我们，不过我真的很喜欢中国的衬衫。现在销量很好。

M: 也许我该开家店。

N: 别忘了在网上也开一个！

Questions 3

1. Do you like to wear Chinese clothes?
2. Do you think that Chinese clothes could become fashionable in the West?

24 常见误解

Common Misconceptions

‖ **Background Information** ‖

There are always common stereotypes of people from different countries and even different parts of the same country. While there is often a lot of truth in them they do not often hold true for all people and for all time.

不同国家，甚至是同一个国家的不同地区，总是会有一些对人们的陈见。尽管这些陈见有一定的道理，但是并不适用所有的人，也不总是正确的。

背景信息

Dialogue 1

Nick: It's a commonly held view in the West that all Chinese look alike.

Rosemary: It's funny you should say that as we think that all Westerners look the same!

N: I've been here in China long enough to see people as individuals but I can't tell which part of China they come from.

R: Well, generally if they are tall and pale then they come from North China. If small and dark they come from South China.

N: What about facial characteristics?

R: If they have a small flat nose they're likely to come from South China.

N: I see. Why do you say all Westerners look alike?

R: Well, you are white, have fair hair and your eyes are blue. All Westerners are like that.

N: Not at all. Many Westerners have brown eyes, grey eyes, green eyes as well as blue eyes. We also have different coloured hair.

R: Such as?

N: Red, blonde, brunette and black.

R: But you are all white, though.

N: We have brown and black skins too. Next time you see a Westerner have a close look!

N: 西方人普遍认为中国人的长相都差不多。

R: 你这么说可真好笑，因为我们也觉得西方人长的都一样。

N: 我在中国待了很多年，所以还不至于看谁都一样。可我分不出来别人是从哪个地方来的。

R: 一般而言，要是一个人又高又白，那就是北方来的；要是又矮又黑，那就是南方来的。

N: 有什么面部特征吗？

R: 鼻子小而平的人，很可能是南方人。

N: 我明白了。你为什么说西方人长的都一样？

R: 白皮肤，金发碧眼，所有的西方人都这样啊。

N: 才不是呢。除了蓝眼睛，西方人也有棕眼睛、灰眼睛、绿眼睛的。头发的颜色也不一样。

R: 比如呢？

N: 红色的、金色的、深色的，还有黑色的。

R: 但你们的皮肤都是白色的啊。

N: 我们也有棕色和黑色的皮肤。你下次再见到老外的时候好好看看。

Questions 1

1. Do you think Westerners all look the same?
2. Can you tell which province Chinese people come from?

Dialogue 2

Nick: Most Chinese seem to think that all Englishmen are gentlemen, drink afternoon tea and that London has a lot of fog.

Rosemary: What's wrong with that? What about fog? I hear that there's a lot in London.

N: Well, that used to be true up to 50 years ago but now London is fog free.

R: Why is that?

N: Ever since the Clean Air Act of the early 50's we have moved away from coal to smokeless fuel and so the air has got a lot cleaner.

R: I've heard that you Englishmen have afternoon tea every day. Is that true?

N: Well, afternoon tea is said to have originated with one person, Anna, 7th Duchess of Bedford. In the early 1800's she

People 人

launched the idea of having tea in the late afternoon to bridge the gap between luncheon and dinner, which in fashionable circles might not be served until 8 o'clock at night. Nowadays, afternoon tea has gone **out of fashion**[1] for the majority of British people but it's very popular among tourists.

R: Don't tell me that Englishmen are not gentlemen!

N: I'm sorry to disillusion you but it's a fact now that we are not so good-mannered or kind to other people.

R: How has that come about?

N: Several reasons. One is the popularity of reality shows like Big Brother where contestants are encouraged to behave in outrageous ways and perhaps one-parent families where there is no father to be a good role model.

R: So we Chinese are becoming more polite while you are getting worse!

N: Seems like it.

习惯用语 2

① out of fashion: 过时的，不流行的

N: 好像很多中国人都觉得英国人很绅士、喝下午茶、伦敦的雾气很严重。

R: 有什么不对吗？对了，你们那里雾怎么样？我听说伦敦的雾很严重。

N: 50 年前是这样，但现在，伦敦已经没有雾了。

R: 为什么？

N: 50 年代初，我们采取了净化空气行动，用无烟燃料代替煤炭，空气因此而变得清洁了许多。

R: 我听说英国人每天都要喝下午茶，是这样吗？

N: 下午茶据说是由一个人发起的，那就是安娜，一位来自贝德福德的公爵。19 世纪早期，她提出用喝下午茶的方式来填补午饭与晚

饭之间的空隙时间，因为那时候在上流时尚界要到晚上 8 点才能吃晚饭。现在，对于大多数英国人来说，下午茶早就不流行了，倒是在游客中很流行。

R: 别跟我说英国人都不绅士啊！

N: 不好意思，我得打破你对英国人的幻想，事实上，我们对别人没有多么有礼貌，或者有多么好。

R: 怎么会变得这样呢？

N: 这有很多原因。其中一个原因就是类似于 Big Brother 这种"真实秀"节目的流行，这个节目鼓励参与者去野蛮地竞争。另外，还有与没有父亲的单亲家庭有关，因为在这样的家庭中，没有父亲来起好的引导作用。

R: 就是说，在中国人变得越来越有礼貌的同时你们却在退步！

N: 好像是这样。

Questions 2

1. Where do you get most of your ideas about English people from?
2. Can you think of any other things about English people that might be out-of-date?

Dialogue 3

Elaine: Oh my! Look at that girl over there, she's very pretty, like most foreigners. But she's very fat.

Tori: Elaine, she isn't fat. Not by foreign standards anyway.

E: What do you mean? She has very large everything! Much bigger than mine!

T: That's because you have a usual Oriental frame, whereas she has a usual Western frame.

E: What do you mean by that? I mean, what is frame like photo frame?

T: No, no. You've got **the wrong end of the stick**[1]. Frame means body type or body shape and size.

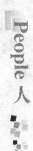

People 人

163

E: Oh, I see. But that doesn't explain why she is so big. Look at her feet! They are bigger than my brother's.

T: Western girls usually are bigger everywhere than Chinese girls. How tall are you Elaine and what shoe size do you take?

E: I'm 155 cm and my feet are a size 37, why?

T: Well, in Western countries you would be considered short. The average height for a girl is around 162–167 cm. And your feet would be pretty tiny.

E: How tall do you think she is? Maybe 180 cm?

T: Not that tall! She's not a basketball player. I would say maybe 165 or 170 cm. Around normal height. And as for her feet, they are certainly not as big as your brother's!

习惯用语 3

① the wrong end of the stick: 得出错误的结论

E: 天哪！你看那边那个女孩，真漂亮，跟大多数老外一样。不过她有点儿胖。

T: Elaine，她可不胖，按照老外的判定标准她真不胖。

E: 那你是什么意思？她个头倒是挺大的，哪儿都比我大。

T: 因为你是典型的东方"骨架"，而她是典型的西方"骨架"。

E: 什么意思？我是指你所说的"frame"是什么意思，它和"photo frame 相框"意思一样吗？

T: 不不，不是那个意思，你理解错了。骨架是指体格类型，身体轮廓大小。

E: 噢，明白了。但你还是没有说为什么她会这么大个儿啊，你看她那双脚，比我哥的还大。

T: 西方女孩一般哪个部位都比中国女孩大。你多高？穿多大的鞋？

E: 身高155厘米，穿37号的鞋，怎么了？

T: 这要是在西方国家，你就是个矮个子了。那里女生的平均身高是162–167厘米。你的脚也简直太袖珍了。

E: 你觉得那个女孩多高？有 180 厘米吗？

T: 没那么高，又不是篮球运动员。我估计她有 165 或者 170 厘米吧。就是一般身高。至于她的脚，肯定不会跟你哥哥的一样大。

Questions 3

1. Would you like to be the same size as a Western girl? Why? Why not?
2. Would you like to be tall and slim or tall and curvy?

25 传统中医
TCM

Background Information

	Western medicine	TCM
Human body	Man is a machine which can be understood by taking it apart to its smallest components	TCM regards man's body as interrelated and constantly interacting with the environment
Illness	Illness is the faulty functioning of various biological mechanisms	Illness is due to disharmony between internal and external factors
Treatment	Treatment corrects the malfunction of a particular mechanism	Treatment activates the body's self-recovery mechanisms

	西方医学	传统中医
人体	人体就像一部机器，可以被单独分解成小的个体来进行研究。	传统中医认为人体与外界环境是彼此相关、相互作用的。
疾病	疾病就是各种生理器官的功能发生了故障。	疾病是由于身体内外因素没有相互协调而引起的。
治疗	治疗的办法就是修护特定的生病器官。	侧重于激发身体的自身修复能力。

背景信息

Dialogue 1

Tom: Hi, Ray, haven't seen you for 3 days. Are you all right?

Ray: I'm OK now, thanks. I've been sick. I went to see my uncle, and stayed at his place for 3 days.

T: What happened to you? I remembered you were **as strong as a horse**[①] when we played football last weekend.

R: Maybe I was trying too hard to be a horse that day. When I got up on Monday morning, **my back was killing me**[②]. So I called my uncle and asked him for advice.

T: I see. I heard your uncle has a clinic here.

R: That's right. He has a TCM clinic, uses traditional Chinese medicine to treat patients. You know, I've had back pain since I finished my report a few weeks ago. Dr. Knowles told me I'd been working in front of the computer too long, and gave me some pain killers.

T: Well, the pain killers work on me very well. What? Did you **run out of**[③] them this time?

R: No. I felt this time the pain was very different from before, so I decided to change the way of treating it. Luckily, my uncle **sorted it out**[④] for me.

T: But how? I heard that traditional Chinese medical treatment is totally different from the medical practice in our world. The TCM doctors use hand, herbs, needles, even five elements to treat patients. What kind of element did your uncle use on you?

R: Fire, this time. Actually, he used cups. Those cups are specially made for medical purposes. He said I'd got bad air in my back, so he had to use the cups to get the air out. Then he just put some fire inside the cups and left them on my back. Apart from the scary marks they left, I'm totally fine.

T: Oh, the cupping method. I remember something about that. You know, once Gwyneth Paltrow was found to have lots of bruises on her body, she claimed that she had had the Chinese cupping method rather than her boyfriend abusing her. I don't know about that, but for me, it sounds like magic, and looks like mistreatment.

R: Shut up, you moron. It looks like I'm a rugby player.

习惯用语 1

① as strong as a horse: 非常健壮，健壮如牛
② my back was killing me: 后背剧痛
③ run out of: 用完
④ sort it out: 解决

T: 嗨，Ray，3 天没见你了，你还好吧？

R: 谢谢，我没事儿。前两天生病了。在我叔叔家待了 3 天。

T: 怎么了？我记得上周末踢足球的时候你还跟一匹小野马似的呢。

R: 可能就是那天踢得太卖力了。周一早上一醒来，后背都快疼死了。于是就给我叔叔打了个电话，问他有什么好办法没有。

T: 哦，我听说你叔叔自己开了个诊所。

R: 没错，他有一个中医诊所，用传统的中药来治疗疾病。你知道，几周前，我完成报告以后就开始觉得后背疼了。Knowles 医生说我是在电脑前工作得太久了，给我开了点儿止疼片。

T: 止疼片对我很有效。你这回治疗效果怎么样？

R: 不怎么样。我觉得这回的疼痛不同于从前，所以我觉得换一种疗法。幸好，我叔叔最终帮我治好了。

T: 怎么治的？我听说传统中医与我们的治疗完全不同。中医用按摩、草药、针灸，甚至五行理论来治病。你叔叔给你用了哪种疗法？

R: 是拔火罐，用一些特制的小罐子。因为他说我的后背受风了，所以他通过使用那些专门用于治疗的小罐子把风拔出去。就是在罐子里面点燃火种，然后把小罐子扣在我的后背上。我现在完全好了，只是后背上留下了一些看起来很吓人的印记。

T: 哦，拔火罐。我记得有一回 Gwyneth Paltrow 身上全是这种淤血的印记，她说是拔了火罐，而不是受到了男朋友的虐待。我对这个不是很了解，但我觉得这有点儿像巫术，看起来有点儿像是虐待。

R: 得了吧，你这个弱智。这让我看起来很像个橄榄球运动员。

Questions 1

1. Have you ever tried TCM?
2. Do you prefer TCM or Western medicine? Why?

Dialogue 2

Troy: I went to hospital yesterday, and found out there is a traditional Chinese medicine department.

Queenie: Yes, there is. But what's wrong with you? Is it serious?

T: Don't worry. I just went to ask for some sleeping pills. You know, you can't get any without prescriptions in a pharmacy. And the doctor told me that I've got slight insomnia.

Q: Actually, you really should have seen a TCM doctor! I think there are many varieties to treat your problem, and less side-effects than sleeping pills.

T: Are you sure? In the West, the TCM clinics are neither included into hospitals, nor in the national medical systems. I know traditional Chinese medicine is an ancient and still very vital holistic system of health and healing, but I need more confidence in it before accepting that.

Q: Well, TCM is based on the notion of harmony and balance, and employing the ideas of moderation and prevention. It is a complete system of health-care with its own unique theories of anatomy, health, and treatment. It emphasizes diet and prevention and using acupuncture, herbal medicine, massage, and exercise; and focuses on stimulating the body's natural

Health 健康

curative powers.

T: It sounds great. How about surgery? Can a TCM doctor do that to save lives in danger?

Q: TCM should not be substituted for contemporary modern trauma practice. Also, TCM is not **the first line of treatment**[①] for bacterial infection or cancer, but maybe usefully complement contemporary medical treatment for those conditions.

T: OK, that means, for example, when Western physicians do surgery and chemotherapy or radiation for cancer patients, TCM physicians might use acupuncture and dietary changes to assist the treatments, right?

Q: You're right. Since the earliest Chinese physicians were also philosophers, their ways of viewing the world and human beings' role in it affected their medicine. Therefore, these philosophical concepts that differ considerably from infection-based principles of medicine and health mean the methods employed by traditional Chinese medicine are also quite different.

T: Ah, that's it. Can I say, if Western practitioners could be described as interventionist and dependent on drugs and other chemical based products, TCM methods are mostly natural and non-invasive?

Q: Yeah, you got it. TCM believes in "curing the root" of a disease and not merely in treating its symptoms. That's very good for your case as well. I think you should go back and see a TCM doctor. And throw those sleeping pills away! They're more trouble than they're worth.

习惯用语 2

① the first line of treatment: 首选的治疗方法

T: 我昨天去医院了，发现那里有个中医科。

Q: 对啊，你生什么病了？严重吗？

T: 没事儿，只是去开点儿安眠药。你知道，没有医生的处方，我在药店里买不到。医生说我有轻微的失眠症。

Q: 其实，你真的应该去看看中医。治你这种病有很多种疗法可以选择，比安眠药的副作用小多了。

T: 你确定？在西方，中医诊所既不会开在医院里，也不在全国的医疗体系内。我知道中医历史悠久，现在依然是一门很重要的健康治疗医学，但要想让我接受它，我还是得再多了解一些，不然没什么信心。

Q: 中医是以调和与平衡为基础，采用调理和预防的理念。以其特有的解剖、康复和治疗理论形成了一套完整的保健体系。它强调饮食与预防，运用针灸、中药、按摩、锻炼来激发人体自身的天然治愈能力。

T: 这听起来不错。那中医能实施外科手术吗？中医能够处理那种致命的外伤吗？

Q: 中医无法取代现在的外伤医疗方法。同时，对于细菌感染、癌症等疾病来说，中医也不是首选的疗法，但可以起到辅助治疗的作用。

T: 哦，那就是说，当西医用手术、射线化疗等方法治疗癌症疾病的时候，中医可以采用针灸、调节饮食等方法来进行辅助治疗，对吗？

Q: 没错。在早期，中医其实也是哲学家，他们对世界以及人类在其中的角色的看法影响了他们的医学。这种哲学观念，与以感染为基础的现代医学保健理念有着相当大的差别，因而中医所采用的治疗方法与西医也有很大的不同。

T: 哦，那我可不可以说，西医是以药物和其他化学药品为基础的强行治疗，而中医则更强调自然疗法，而非强行治疗？

Q: 是的，没错。中医讲究的是对疾病要"去根"，而不是"治表"。这对你的病来说就很好。我觉得你应该回去看看中医，把那些安眠药扔掉吧，它们只是花钱找罪受，不值。

Dialogue 3

Troy: Yesterday, my doctor was so serious with me, that I was a little afraid. I just wanted some sleeping pills, but he suggested that I do a full body check, even scan my head!

Queenie: He's right. This doctor is very analytical and responsible. He doesn't want to delay your disease if there is a problem. Of course, "if" is the word here, hopefully there is nothing wrong with you. In Western medicine, patients with similar complaints or diseases usually will receive virtually the same treatment.

T: How about Traditional Chinese Medicine diagnosis? Does every patient get treated a different way?

Q: Well, in TCM, the physician treats the patient rather than the condition, and believes that identical diseases can have entirely different causes. TCM offers a more humane, patient-oriented approach that encourages a high degree of practitioner-patient interaction and is not overly dependent on technology.

T: OK. What will the TCM doctor do? Is there any special type of examinations? Or should I take some time to interact with him, so we can get to know each other? Actually, I'll probably have to ask you to come with me, because of my poor Chinese.

Q: During the important first visit, the practitioner will conduct four types of examinations, all extremely observational. First, the practitioner will ask many questions, going beyond the typical patient history to inquire about such particulars as eating and bowel habits or sleep patterns.

T: That's not the same as my doctor. He asked me a few questions about head injury, stress, getting a cold, and so on.

Q: OK. Next, the doctor will look at the patient, observing his or her complexion and eyes, while also examining the tongue very closely, believing that it is a barometer of the body's health and that different areas of the tongue can reflect the functioning of different body organs.

T: Really? Can TCM doctors define the problems through eyes and tongues? I know they touch the patients' wrists and check the pulse, and then stab needles in somewhere on the body, where they believe it will be effective!

Q: Oh, come on. Let me finish. They will check the patients' pulse, after listening to the patients' voice or cough, and smelling breath and odours. Through six different pulses, the well-trained doctors can diagnose any problem with the flow of the all-important Qi. In general, these essential examinations will lead the physician to diagnose or decide the patients' problems.

T: This diagnosis is very different from one in contemporary Western medicine. No blood or urine samples are tested in a laboratory. No brain scan is needed. I think that's great for my problem. I'm scared of those big machines that scan your head and I've never enjoyed peeing in a cup.

Q: Yeah, you're right. However, the key to this technique lies in the experience and skill of the practitioners.

T: 昨天医生的那个严肃劲儿真是吓到我了。我只是想开点儿安眠药，可他却让我做个全身检查，甚至做了脑部扫描。

Q: 他这样做是对的，这个大夫善于分析且很有责任心。他只是不想延误你的病情，万一有什么状况怎么办。当然，也只是个万一，希望你一切正常。在西医里，如果患者有类似的症状，会接受同样的检查。

T: 那中医如何诊断？每个病人都会有不同的治疗方法吗？

Q: 中医更加重视病人本身，而不是疾病的症状，他们相信即使是同样的疾病，其诱因也不尽相同。中医会采用更加人性化的、以病人为导向的治疗方法，它强调的是医生与病人之间高度的互动交流合作，而不是过分依赖现代科技。

T: 那中医做什么？有什么特殊的检查方法吗？我是不是应该与他们多互动交流来增进彼此的了解？我中文不好，得带上你跟我一起去。

Q: 第一次看中医很重要，医生会进行四项检查，最重要的就是观察。首先医生会问你一些问题，除了疾病本身外，还会事无巨细地问你其他的情况，比如你的饮食、大便情况、睡眠等等。

T: 这可与西医不一样。他问我的头疼状况、压力情况、有没有着凉等等。

Q: 嗯，然后医生会对你进行观察，观察你的面色、眼睛，仔细地检查你的舌头。他们相信舌头就相当于人体健康状况的晴雨表，而且舌头的不同部位，可以反映出不同身体器官的相应功能。

T: 真的？中医从眼睛和舌头就能找到问题所在？我知道他们会握着病人的手腕来检查脉搏，然后在那些他们认为会有疗效的身体部位进行针刺。

Q: 哦，得了，我还没说完呢。他们先聆听病人的声音或者咳嗽声，闻病人呼入呼出的气味，然后再检查病人的脉搏。根据六种不同的脉象，经验丰富的医生可以诊断出人体内的生命之源——气的各种问题。这些重要的检查可以帮助大夫找到病人的问题所在。

T: 这种诊断与当代西方医学的诊断大不相同。不需要对血和尿进行化验，也不需要进行脑扫描。对我的病来说这是最好的办法了，我讨厌那些对大脑进行扫描的大机器，也不喜欢在杯子里排尿做尿检。

Q: 没错。不过，中医的诊断需要医生的经验和技术。

Questions 3

1. If a TCM doctor felt your pulse and told you what your problem was would you believe him?
2. Do you think doctors should be trained in both TCM and Western medicine? Why? Why not?

26　针灸

Acupuncture

Background Information

　　Acupuncture, one of the main forms of therapy in traditional Chinese medicine（TCM）, has been practiced for at least 2,500 years. In acupuncture, certain points on the body are stimulated by the insertion of fine needles. Unlike the hollow hypodermic needles used in mainstream medicine to give injections or to draw blood, acupuncture needles are solid. The points can be needled between 15° and 90° relative to the skin's surface, depending on treatment.

　　Acupuncture is thought to restore health by removing energy imbalances and blockages in the body.

　　针灸是传统中医的主要疗法之一，至今已经延续使用了2500多年。在针灸治疗中，采用细针灸在身体的某个部位起到刺激的作用。与注射药品和抽血所使用的空心针头不同，针灸疗法所使用的针是实心的。根据皮肤表面状况和治疗的需要，针头会与皮肤表面成15°至90°角。

　　针灸疗法，被认为是通过除去体内能量失衡和阻塞来恢复健康的。

背景信息

Dialogue 1

Tori: Hi, Ray. Yesterday I walked past a TCM clinic, Dr. Wang's Clinic. Is Dr. Wang your uncle?

Ray: Yes. My uncle is Dr. Wang. What's up?

T: Well, when I passed the shop, I noticed the poster on the window, which says there is a new way to quit smoking and drinking. I want to know more about that.

R: Wow, I didn't realise you had these kinds of problems.

T: Of course, it's not for me. My dad has wanted to quit smoking for ages, he's tried many ways, but in the end, he always started up again. I wondered if your uncle can help him.

R: I see. Well, this treatment is based on the acupuncture theory. You know, a Chinese doctor uses special needles to sting certain points of your body to cure your illness. In this treatment, the pressure points are all on your ears, stimulating these points can numb your desire for smoking.

T: All right, I got it now. But I'm a little confused about piercing ears with needles, won't they bleed?

R: Oh, sorry. I didn't mention that the doctor uses special seeds instead of needles. The seeds will be fixed individually by little pieces of plaster on the points on your ears. Gently pressing the seeds for a while many times a day will stop you thinking of cigarettes.

T: Is this treatment popular in China?

R: Yes. But in China, ear acupuncture is used mostly to treat the problem of short sightedness and unknown pain. Nowadays, it's been extended into weight loss, for quitting drinking and smoking, even treating snoring.

T: Well, thank you so much for that useful information.

Especially, there are no needles in ears. It makes me think about that guy, Pin Head in the film *Hell Raiser*! Too scary for me!

R: We are not trying to scare people to death, if you don't quit smoking that'll kill you anyway, with no help from us.

T: 嗨，Ray，我昨天路过了一个"王医生中医诊所"。是你叔叔开的吗？

R: 是的，我叔叔就是王医生。

T: 我路过诊所的时候，看到窗户的海报上说有一种戒烟忌酒的新疗法。我想了解一下。

R: 哇！我怎么没发现你还有抽烟喝酒的习惯啊。

T: 当然不是我了，我爸爸想戒烟已经好多年了，尝试了很多办法都没效果。我想看看你叔叔的办法是不是有效。

R: 我明白了。他的治疗方法是以针灸理论为基础的。你知道，中医用特制的针刺入身体来治病。他的这种方法是要刺在耳朵上，刺激穴位，阻止你的吸烟欲望。

T: 哦，我明白了。但用针刺耳朵，我有点儿疑虑，不会出血吗？

R: 哦，对不起，是我没说清楚，医生用的是耳针，就是类似于小种子那样，而不是针灸用的针。这些小种子被固定在单独的胶布上，然后贴在耳朵上相应的穴位处。每天轻轻按压几次耳针，就不想吸烟了。

T: 这种疗法在中国很流行吗？

R: 是的。但是在中国，耳部针灸大多用来治疗近视和身体的疼痛。不过，现在也可以用来减肥、戒烟戒酒，甚至用来治疗打呼噜。

T: 谢谢你提供的信息。尤其是不在耳朵上扎针。这让我想起了 Pin Head，就是《养鬼吃人》里面那个家伙，吓死我了!?

R: 我们可不想把人吓死，要是不戒烟的话就会因此丧命，连我们也没办法。

Questions 1

1. Have you ever tried acupuncture? Why? Why not?
2. Do you think acupuncture works? Why? Why not?

Dialogue 2

Ray: Hi, Tom. What are you doing here?

Tom: That's what I want to ask you! I'm on my way home, just finished my shift.

R: I sometimes work here. I'm his favourite nephew, if my uncle gets **short-handed**① at weekends, I'm always the first to be chosen.

T: Wow, you really must love each other.

R: I'm joking. I like to work here, I'm not that busy anyway, and I'll learn the basic knowledge of herbs and Chinese medicine.

T: Good for you. You know, one of my cousins will go to China to learn traditional Chinese medicine later this year.

R: Well, good for him then. Actually, some universities here offer the TCM degrees as well. Anyway, look at that poster on the wall. These are all the acupuncture points on our bodies. They all relate to our organs, tissues, and nerves. Some of them are lethal.

T: Oh, I know that. In some Kung Fu movies, Ninjas can kill enemies or disable them by hitting some points on their bodies.

R: Come on, we're talking about Chinese medicine and saving lives. When you start learning acupuncture, you should memorise the pressure points and the inside parts related to them. Then you'll learn how to control the needles. It could take your whole life to learn that. And unless you're fully trained, you're not allowed to get even close to those lethal points.

T: Wow, it sounds very serious. But we regard acupuncture and other Chinese medical treatments more as cultural attractions than as life preserving.

R: Well, that will explain the fact that many Western countries lifted the ban on traditional Chinese medical clinics a couple of decades ago.

T: OK. We never banned TCM clinics, we just don't include them into the NHS? So you must pay for the treatment.

习惯用语 2

① short-handed: 人手不足的

R: 嗨 Tom，你在这儿干嘛呢？

T: 我正想问你这个呢。我要回家啊，刚刚值完班。

R: 我有时候在我叔叔的诊所帮忙。要是我叔叔周末缺人手的话，我就是第一人选，因为我是他最爱的侄子。

T: 哇！你们爷俩感情真不错啊。

R: 开个玩笑。我很喜欢在这里工作，再说我也不忙，还可以学一学草药和中医的基本知识。

T: 这倒还不错。你知道，我的一个堂兄下半年要去中国学中医。

R: 这很好啊，其实这里的一些大学就能读取中医学位。你看墙上的那张海报，都是人体的穴位图，这些穴位与我们的器官、组织和神经有联系。有些穴位是致命的。

T: 哦，我知道，在一些功夫电影里，武士们可以通过点穴来杀死敌人或使他们不能动弹。

R: 嘿！我们正在讨论用中医治病救人的话题，你扯哪儿去了。学习针灸的时候，得记住穴位及其对应的人体部位。然后再学如何控制用针。用一生的时间也学不完，除非你得到了完整的培训，否则不允许你接近那些致命的穴位。

T: 哇！这么严重啊，我们更多的是把中医疗法视作一种文化，而不是治病的方法。

R: 所以这就是为什么二三十年前中医诊所在很多国家都已解禁。

T: 我们可从没有禁止中医诊所，我们只是把它排除在医保体系之外。所以，要想用中医进行治疗的话，你就得自己掏腰包。

Questions 2

1. Is there a TCM department in your university?
2. Are there many international students who come to study it?

Dialogue 3

Tori: Thank you so much for having recommended me the traditional Chinese medicine treatment. The only thing that embarrasses me is I can cope with the herbal medicines, but I'm still not quite sure about the acupuncture treatment. You know, I'm scared by needles. And in this case there are lots of them!

Quinn: Don't worry. Lots of people have the same feelings as you. The bunch of needles makes me nervous too, but really, there's nothing to worry about. You won't feel a thing.

T: But why did the pioneer Chinese practitioners invent this painful treatment?

Q: Well, first of all, like I said, it's less painful than you think. Secondly, thousands of years of practice prove that it works very well. It is often used for pain relief, but has wider applications in traditional Chinese practice, such as disorders of the internal organs as well as muscular and skin problems.

T: Really? I know that acupuncture is probably the form of treatment most familiar to Westerners. But how does it work? Is it based on anatomy as well?

Q: No. Acupuncture is based on the notion that the body's vital energy force, Qi, travels through known channels or "meridians". Disease is attributed to a blockage of the meridians; therefore, the insertion of needles at specific points along the meridians is thought to unblock the Qi.

Health 健康

T: Oh, it's just like if you get an embolism in one of your veins, you need to have surgery to get the thrombus out. In this situation, it's needles instead of a knife. Am I right?

Q: Kind of. Actually, the meridians are intangible. Apparently, there are fourteen channels and numerous collaterals under the body surface, which connect the body surface to various internal organs. Along them are more than 800 acupuncture points. However, no contemporary scientific explanation exists as to how or why acupuncture works.

T: So, you mean there is no scientific consensus as to whether acupuncture is effective or only has value as a placebo. But according to the protocols of evidence-based medicine, it's not too difficult to conduct reviews of existing clinical trials.

Q: Yes, some institutes did the reviews, then they fell into disagreement on what is acceptable evidence and on how to interpret it, but further investigation is needed. Anyway, the reviews showed some patients report immediate improvement, others feel exhilarated, while some feel like sleeping. In some cases, patients say their condition worsens before it improves.

T: Well, nonetheless, it has proven effective in treating many types of aches and pains and in helping people with depression and fatigue. But I'm considering the small chance of not only getting an infection from acupuncture, but also that an existing infection could be spread to other parts of the body by increased blood flow and circulation. Maybe I'm just paranoid!

Q: Nowadays, there are many alternatives, such as magnets, mild electric currents, manual pressure, known as acupressure, or even low frequency lasers. All of these can also stimulate these acupuncture points to the same effect. So if you're too scared, try another way.

T: 谢谢你给我推荐中医疗法，但是有一件事让我很尴尬，那就是我能接受中药，可还不太能接受针灸疗法。你知道，我很害怕那些针，而针灸疗法里却偏偏有很多针。

Q: 别担心，很多人都跟你有同样的感受。这一连串的针也让我紧张。但其实没什么可怕的，你不会有任何感觉。

T: 中医的鼻祖们为什么会发明这种略带痛苦的治疗方法啊？

Q: 首先，就像我所说的，针灸并不像你想象的那么疼。其次，几千年的实践证明了针灸很有效。它常被用于减轻疼痛，而且它在中医里还有更广泛的应用，比如解决内脏器官的问题，肌肉的问题以及皮肤问题。

T: 真的吗？我知道针灸大概是西方人最熟悉的中医疗法之一。但它的原理是什么？也是以解剖学为基础吗？

Q: 不是。针灸是基于这样一种理念：认为"气"是人体的重要能量来源，它在体内是沿着经脉运行的。生病的原因是因为经脉阻塞，因此，沿着经脉在人体某些穴位下针，就可以将气打通。

T: 哦，就好像你的血管得了血栓，你就得通过手术将血栓取出一样。在这种情况下，中医就是用针代替了刀，对吗？

Q: 差不多。其实，经脉是无形的。在皮肤下面，共有十四条主脉络，和众多副经脉连接着人体的各个器官。沿着经脉有 800 多个针灸穴位。但是现代科学也无法解释针灸的原理是什么，为什么针灸有疗效。

T: 你是说，并没有科学依据来证明针灸究竟是确实有效，还是仅仅是一种对患者的安慰。但是根据有科学依据的医学原理，对现有的临床试验进行评审并不困难啊。

Q: 的确有研究机构做过这样的评审。但是他们随后陷入这样的分歧：究竟什么才算是可接受的证据，以及如何解释这些证据。因而仍需要进一步调查研究。不管怎样，这些评审的结果显示，有些患者在接受针灸治疗后，的确有立竿见影的疗效，有些则感觉兴奋，还有人感觉昏昏欲睡。当然也有病人说在接受针灸治疗之后，病情反倒恶化了。

T: 但是，尽管如此，现在已经证明，针灸在治疗很多病痛以及缓解疲乏方面成效显著。不过，我也在考虑这样一种可能，尽管可能性不大——针灸不仅会受到感染，而且使得现有的体内感染加速

Health 健康

扩散到身体其他部位，因为针灸会加速体内的血液循环。当然，也许是我过分怀疑了。

Q: 现在还有许多其他的针灸疗法，而不一定非得用针，比如有用磁体的，有用轻微电流的，有用手压的叫做手压法，甚至还有用低频激光的，所有这些针灸疗法对针灸穴位的刺激效果相同。如果你真的如此担心，就试试别的。

Questions 3

1. If scientists can explain acupuncture would you try it? Why? Why not?
2. Have you ever tried any unorthodox medical treatment?

27 风水
Feng Shui

‖ Background Information ‖

Feng Shui is the ancient Chinese art of arranging buildings, objects, and space in the environment in order to achieve energy, harmony, and balance.

It is also known as geomancy. It is heavily influenced by the theory of the five natural elements of metal, wood, water, fire and earth. These five elements either complement or check each other and a Feng Shui expert will make suggestions on how to achieve balance or harmony.

风水是一种古老的中国艺术，它是对建筑、物品、环境空间布局进行合理安排，以便达到能量、和谐与平衡。

风水也叫土地占卜，它深受五行（金、木、水、火、土）理论的影响。据说，这五行互相补充或制约，而风水大师能够就如何取得五行平衡或和谐给予建议。

背景信息

Dialogue 1

Rose: Nice party, Tom.

Tom: Hi, Rose. I'm glad you enjoy the party. What do think about this place? You haven't been here since we redecorated it.

R: Um, to be honest, I feel it's a little bit strange, looks like some kind of East-meets-West style, but in a good way. I believe you've put a lot of effort into it.

T: You're right. My new house mate is obsessed with Feng Shui. Since he moved in, he has spent lots of time and energy on changing things, rearranging furniture, and redecorating. I've helped him.

R: Did you learn anything from him? I mean Feng Shui stuff.

T: Not quite, yet. I just know Feng Shui is the way of Wind and Water or the natural forces of the universe. And good Feng Shui is believed to promote health, prosperity, creativity, positive social relationships, self-confidence, contemplation, and respect for others.

R: Wow, amazing. You've grasped the core of Feng Shui. Do you believe the theories?

T: I really don't know. He told me absolutely lots of stuff, but I only can remember something like "Yin and Yang", "Qi and Xiang", "high and low", "side and middle". Anyway, I feel all the work we've done makes this house more comfortable. So, I guess I can accept these concepts.

R: Well, I don't think I know more than you. I don't know how to apply the Feng Shui theories in real life either.

T: Um, from this point of view, Feng Shui is not strictly a Chinese entity. I read that in prehistoric Europe, the practice of arranging objects and structures to be in harmony with the universe was a relatively common practice.

R: Ah-ha, somebody did some background checks? You seem to know everything. By the way, did he mention which room is the best place to have a party in this house?

T: I don't think so. He went out of town to visit his parents. He just told me not to touch his goldfish after we finished the house.

R: Tom，派对真不错。

T: 嗨 Rose。很高兴你能来参加这个派对。觉得这个地方怎么样？自从我们重新装修以后你就一直没来过。

R: 嗯，老实说，我感觉这里有点儿奇特，有种东西交融的感觉，但却很不错。你肯定是花了很多心思在这上面。

T: 没错，我的新室友对风水很有研究。自从他搬进来后，花了很多时间和精力重新布局家具，装修房子。我也一直在帮他。

R: 你没学两招儿？我是说风水方面的事。

T: 还没有。我只知道风水是关于风和水或者宇宙自然力量的东西。人们通常认为，风水好有助于健康、发财、兴旺，以及建立积极的社会关系，提升自信，有好的前景，而且对别人也有好处。

R: 哇！太神奇了！你算是掌握了风水的精髓。你相信这个吗？

T: 我还真不知道，他是跟我说了很多，但我只记得一些，比如阴和阳、气和象、高和低、偏和中。反正我觉得我们做的这些工作让屋子看起来更舒服了。所以，我想我能接受这些理念。

R: 我知道的比你少多了，而且也不知道如何把风水理论应用到实际生活当中。

T: 嗯，从这一点来讲，风水其实并不是中国特有的，我读过资料，说在史前的欧洲，人们摆放物品，布局结构，以使之与宇宙相互和谐，这在当时也是非常讲究的。

R: 哈哈，看来你还真是做过背景调查嘛。你好像什么都知道。还有，他有没有说这个房子里哪个房间最适合办派对？

T: 没说。他去外地看望父母了。他只是告诉我派对结束后，不要碰他的金鱼。

Questions 1

1. Is your flat arranged in accordance with the principles of Feng Shui?
2. Do you keep any goldfish?

Dialogue 2

Frank: Hello, Rose, I'm Frank, Tom's house mate.

Rose: Hi Frank, nice to meet you. What can I help you with?

F: Well, I've been learning Feng Shui recently and I'm a little confused. Tom told me you know lots about Feng Shui, so I wonder if I can ask you something.

R: I've been studying basic Chinese medicine, so I just know a little bit about Qi and Ying Yang. But their functions are not quite the same as those in Feng Shui.

F: Oh, really? But it doesn't matter. I read that in recent years, many Feng Shui books have been published in English, and often focused on interior design, architecture, and landscape design.

R: Yes, it sounds like Feng Shui studies in the Western world are more practical than theoretical.

F: That's what I'm learning now. I also learned that it is unclear what relationship these Western interpretations of Feng Shui have to the Eastern tradition. Moreover, Asian Feng Shui practitioners regard Western adaptations as inauthentic. So I want to know some original foundations.

R: I see.

F: 你好 Rose，我是 Frank，Tom 的室友。

R: 你好 Frank，很高兴见到你。有什么需要我帮忙的吗？

F: 我最近在研究风水，有点儿困惑。Tom 说你对风水很有研究，所以我想跟你请教些问题，可以吗？

R: 我学过一些中医基础，所以知道一些关于气和阴阳的东西。但它们的功能与在风水中不太相似。

F: 真的？不过没关系。我看资料说，在近些年出版了很多英文版的风水书籍，这些书籍往往是集中在风水与室内设计、建筑和景观设计等方面的关系。

R: 听起来，西方的风水研究更加侧重于实践啊。

F: 我现在就在学这个，而且也了解到，西方对风水的解释与东方的风水传统之间有什么关系，这目前还不清楚。而且，亚洲的风水大师们认为西方对风水的改编，使风水变得不再正宗。所以我想知道一些关于风水方面的本原的东西。

R: 我明白了。

Questions 2

1. Can Feng Shui be adapted in a Western way? Why? Why not?
2. Is it right for some Feng Shui principles to be adapted in this way or do you think all of it should? Why? Why not?

Dialogue 3

Rose: You know, last week, I **googled**[①] Feng Shui, and found a few interesting facts. For example, Donald Trump and Prince Charles are said to have used Feng Shui.

Frank: That's not even a secret! The *Los Angeles Times* reported that News Corp., Coca-Cola, Proctor & Gamble, Hewlett-Packard and Ford Motors are also using Feng Shui.

R: Wow, they're all big names. In mainland China, nowadays, Feng Shui is regarded as a form of superstition, and dismissed as pseudoscience. But in the West, it looks like Feng Shui has found its new territory.

F: Oh, not quite like that. In the West, most people have reacted sceptically towards the purported benefits of crystals, wind chimes, table fountains, and mirrored balls, on one's life,

Health 健康

finances, and relationships. Often, these claims are rejected as New Age, relying on the placebo effect, or even outright fraud.

R: That's strange. Why would celebrities and commercial giants accept that?

F: Yeah, maybe they just believe that some of its more practical rules are very useful, such as not working with one's back to a door, even some extreme critics said it's a mixture of rough guesses at nature and fanciful play with silly diagrams.

R: I think it's maybe because famous people like to believe in something. If you believe that something can't hurt you, it won't.

F: Yeah, maybe. But I think Feng Shui is quite useful in interior design, architecture, and landscape design.

R: I think so. The goal of Feng Shui is to locate and orient dwellings, possessions, land and landscaping, so as to be tuned in with the flow of life force or spiritual energy.

F: That's right. Apparently, location is considered to be of far greater significance than orientation, according to the Feng Shui guideline. This is in line with modern thinking, where the 3 principles of buying a piece of property are location, location and location!

习惯用语 3

① googled: 在 Google 网上进行搜索

R: 上周我在 google 网上搜索"风水"一词时，发现了很多有意思的事儿。据说唐纳德·邓普（美国著名的地产大亨，美国著名电视节目"学徒"（The Apprentice）就是由他制作拍摄的——译者注）和查尔斯王子就运用过风水理论。

F: 这也不是什么秘密啊！《洛杉矶时报》说新闻集团、可口可乐、宝洁公司、惠普、福特汽车都运用风水理论。

R: 哇！可都是大公司啊！现在在中国大陆，风水被认为是一种迷信，被斥为是伪科学。但在西方，风水是一个全新的领域。

F: 哦，也不全是那样。在西方，有人认为水晶、风铃、桌上喷泉、玻璃球据说对于人们的生活、财富和人际关系都有着好的影响，但是大多数人对此还是持怀疑态度。通常，人们将这些奇谈怪论斥为"新时代"理论，这些主要是帮人们寻求一种心理安慰，或者纯粹是欺骗。

R: 那就奇怪了，为什么会有那么多名人和商业大公司接受它呢？

F: 是啊，可能是因为他们觉得这些理论中的有些实用规则还是比较有用的，比如工作时不要背对着门。不过，甚至有一些极端的批评家认为，这套理论只不过是对自然界胡编滥造式的猜测，加上胡乱地摆弄那些愚蠢的图形而已。

R: 我倒是觉得，这可能是因为名人们都喜欢相信点儿什么的缘故，他们或许觉得。如果你相信某东西不会伤害到你，那么它就真的不会伤害你。

F: 也许吧。不过我认为，风水理论在室内设计、建筑以及景观设计方面还是有用的。

R: 我也这么觉得。因为风水理论的目标就是安排好住所、土地以及园林景观的位置和朝向，以便与自然力量或者某种精神能量和谐一致。

F: 没错，按照风水理论，很显然，位置要比朝向重要得多。这就与现代人们的思想观点不谋而合了，比如买房的时候，人们考虑的三大原则是：位置、位置，还是位置！

Questions 3

1. Do you know of any celebs who use Feng Shui?
2. After reading about Feng Shui will you adopt some of its principles? Why? Why not?

Health 健康

28 阴阳
Yin and Yang

Background Information

The two concepts Yin and Yang or the single concept Yin-Yang originate in ancient Chinese philosophy and metaphysics, which describe two primal opposing but complementary principles said to be found in all objects and processes in the universe.

Yin literally "shady place, north slope (hill), south bank (river); cloudy, overcast" is the darker element; it is passive, dark, feminine, downward-seeking, and corresponds to the night. Yin is often symbolized by water or earth, while Yang is symbolized by fire, or wind. Yin (receptive, feminine, dark, passive force) and Yang (creative, masculine, bright, active force) are descriptions of complementary opposites rather than absolutes.

阴和阳这两个概念，或者说阴阳这一对概念，源于中国古代的哲学和形而上学。讲的是既相互对立又相互补充的两个基本原则，据说宇宙万事万物当中都存在这两大原则。阴，字面意思就是阴暗的地方，山的北面，水的南面，也指阴云的天气。它代表了事物黑暗的一面，表示的是被动、阴暗、女性、衰落，对应于夜晚。水或地球往往象征着阴，而火或风则象征着阳。阴（顺从接受的、女性的、阴暗的、被动的力量）和阳（有创造力的、男性阳刚的、明亮的、积极主动的力量）描述事物相对的一面，而不是绝对的一面。

背景信息

Dialogue 1

Nick: What does that black and white symbol mean?

Quinn: It stands for Yin and Yang. You can see that there are similarities in the shape but yet they are opposites and also complementary.

N: In the West we have good and evil. They are opposites but not complementary.

Q: Right, but your good and evil are always at war while Yin and Yang seek harmony.

N: We believe that harmony and peace can only exist when evil has been defeated by good and no longer exists.

Q: That is the difference between our two philosophies. Yin and Yang are equal forces.

N: We think that good is represented by God and bad by the Devil. God is stronger than the Devil and so will triumph in the end.

Q: Well, we believe in **the here and now**[①], not in some far distant future.

N: Yes, but you regard them as forces that need each other and cannot exist without one another while we see them as personifications and not equal.

Q: That's right. Yin and Yang have always existed and once man is at harmony with himself then he can be in harmony with the rest of the world.

N: I agree with the last part. We both seek harmony which you can achieve by yourself but Christians believe that peace can only be achieved by God working within us.

Q: Two very different philosophies that can never be reconciled but with the same end in mind.

Health 健康

N: 这里的黑和白符号分别表示什么意思啊？

Q: 它代表了阴和阳。你可以看到，它们形状相似，但互为对立也互为补充。

N: 在西方，我们说善和恶，它们是相互对立，但不相互补充。

Q: 是的，但你们说的善恶总是矛盾对立的，而我们的阴阳是追求和谐统一的。

N: 我们认为，和谐与和平必然是在恶被善征服之后才能达到。

Q: 这就是东西方两派哲学思想的差别所在。我们认为阴和阳是平等的力量。

N: 我们认为，上帝代表了善，而魔鬼代表了恶。上帝比魔鬼强大，所以善终究会战胜恶。

Q: 噢，我们相信的是此时此地的当前，而不是遥远的未来。

N: 是的，你们把阴和阳看作是你中有我、我中有你的两个力量，彼此共生，缺一不可；而我们把善与恶都赋予了具体的化身，而且把它们看作是不平等的。

Q: 你说的有道理，所以阴和阳总是一直共生共栖，而且一旦一个人找到了自己内心的和谐，他就能够与世界和谐共处了。

N: 我同意你后半部分的观点。我们都是寻求和谐，你们认为通过自身可以达到和谐，但是基督徒认为只有上帝与我们同在才能达到和谐。

Q: 看来，东西方两派哲学思想，虽然彼此有分歧，但最终目的都是一样的。

Questions 1

1. Can you explain Yin and Yang?
2. Do you have a philosophy that explains the world?

Dialogue 2

Tori: I heard people always talk about Yin and Yang, especially, when they are talking about health. So what are Yin and Yang?

Quinn: Well. According to ancient Chinese philosophy, two basic principles underlie all matter and energy in the universe, namely Yin and Yang. These forces are opposites, but are not in opposition.

T: Sorry, I'm a bit confused now. What do you mean?

Q: The theory of Yin and Yang derives from Taoism. In Taoism, the world is full of opposites: anything beautiful and ugly, small and big, good and evil, young and old, male and female, alive and dead, etc. Yet they depend on each other; without ugliness, there is no beauty; without evil, there is no good.

T: Ah, I would say they are complementary. So can you give me some examples?

Q: OK. Yin signifies earth, passive, negative, female, yielding, or dark; Yang signifies heaven, active, positive, male, strong, or light. These principles can be seen throughout nature and in the human body.

T: I see. The human body is so complicated, but by using Yin and Yang theory, different health conditions can be described. They relate to mental, physical, and spiritual structure and are affected by food, drink, action, and inaction.

Q: Yes. When you are in good health you keep a good balance of Yin and Yang. Once you lose the balance, you fall ill.

T: Really? How can I keep a good balance? Are Yin and Yang held in absolute stasis or in movement?

Q: Well, since all forces in nature can be seen as having Yin and Yang states, they are more or less fluctuating. The constantly changing interactions of Yin and Yang give rise to the infinite variety of patterns in life. Homeostasis is the only way to keep Yin and Yang in a balance.

T: All right. I sort of understand now. For instance, female is Yin and male is Yang, they need each other to exist and flourish. The ratio of female and male in the world is homeostasis.

Health 健康

Q: Wow, you are a fast learner! And you explained that in a way that's easy to understand.

T: 我常常听到人们在谈论阴阳，尤其是当他们谈及健康的时候。那么到底什么是阴阳？

Q: 噢，这个问题啊。根据中国古代的哲学思想，宇宙万事万物之间都存在两大基本原则，即阴和阳。这两派力量相互对立，但并不排斥。

T: 对不起，我现在有点儿糊涂了。你说的是什么意思？

Q: 阴阳的理论源自道教。道教的观点认为，世间万物都分阴阳两面，比如美与丑、小与大、善与恶、少与老、男与女、生与死，等等。它们相互对立，但又相互依存。比如，没有丑，就没有美；没有恶，就没有善。

T: 是啊，我觉得它们是互为补充。你能给我举些例子吗？

Q: 好的。阴意味着地球、被动、负面消极、女性、顺从或阴暗，而阳则表示上天、主动、正面积极、男性、强壮或阳光。这些对立面在整个自然宇宙，以及我们人体中都有体现。

T: 我明白了。人体是如此复杂，不过借助于阴阳理论，我们就可以描述不同的健康状况，比如有心理方面的、肉体方面的，还有精神方面的健康，它们都受饮食、活动与否的影响。

Q: 是的，如果你身体健康，那么你就是处于阴阳平衡。若阴阳不平衡，你就会生病。

T: 真的？那我如何才能保持好的阴阳平衡呢？阴阳平衡是静止不变的还是动态变化的？

Q: 因为世间万物的力量都被看作有阴阳两面，所以阴阳平衡是变化波动的。阴阳之间永远是互动作用，这就产生了无数的生命形态。阴阳之间的平衡永远处于一种动态平衡的状态。

T: 好的，我现在有点儿懂了。比如，女性是阴，男性为阳，他们互相需要，彼此共荣。而男女比例也是动态平衡的。

Q: 哦，你学得倒是挺快啊。你这种解释倒是非常容易理解。

Questions 2

1. Do you think Yin and Yang affect health?
2. In China today there are more boys than girls so is there an inbalance there and what should we do?

Dialogue 3

Tori: Are you all right? You look terrible.

Quinn: I'm sick. I've got a headache and toothache. I think I've got fire inside. Recently, I was so busy, so I can't get a good rest, and also I drank too much and ate a lot of meat and seafood. So I kept the fire inside.

T: What do you mean? How can you get or keep fire inside your body? Are you sure you didn't get some kind of infection?

Q: Actually, I mean I've got too much Yang inside. I broke the homeostasis of Yin and Yang. The headache and toothache are on the same side. I just need some good rest, herbal tea, and light food.

T: I have no idea what you are talking about. I know some Yin and Yang, but I still can't understand how you broke the balance of them.

Q: Traditional Chinese physicians built up an understanding of the location and functions of the major organs, and then matched them with the principles of Yin and yang. For example, the liver is Yin and the gall bladder is Yang; the heart is Yin and the small intestine is Yang; spleen is Yin and the stomach is Yang; the lungs are Yin and the intestine is Yang; and the kidneys are Yin and the bladder is Yang.

T: Wait. There is too much information within such a short time. Please give me a simple explanation. I don't get it. Basically, I just want to know how you can keep the balance of Yin and Yang.

Q: OK. There are some rules of Yin and Yang: they can be further subdivided into big Yin, little Yin and big Yang, little Yang; they consume and support each other; they can transform into one another; and part of Yin is in Yang and part

Health 健康

197 《《《

of Yang is in Yin.

T: Oh, it's still quite complicated. Anyway, I sort of understand what you said. Could you please analyse your own system by applying all those theories.

Q: Yeah, that's the easy way to let you know. I didn't sleep enough, so I got Yin deficiency and excess Yang; alcohol is Yang, so my liver turned into too much Yang, too much meat and seafood need my gall bladder to be involved more than normal, too much Yang again!

T: Well, you sound right. But how can you explain the connection between too much Yang and your body aches?

Q: Chinese doctors viewed the body as regulated by a network of energy pathways called meridians that link and balance the various organs. One of the meridian functions is to connect the internal organs with the exterior of the body. So, when inside organs got fire, a tooth ached.

T: 你怎么了？怎么气色看起来这么不好。

Q: 我病了。头痛，牙也痛，感觉可能上火了。最近太忙了，所以没有休息好，而且也有些暴饮暴食，吃了很多肉和海鲜，所以体内有火。

T: 什么意思啊？你体内怎么可能有火呢？可能是疾病感染了吧？

Q: 我的意思是说我体内阳气太旺了，打破了阴阳平衡，所以同时出现头痛和牙痛。我只需要好好休息，喝点儿中药茶，而且饮食清淡些。

T: 我不懂你在说什么。我了解一些阴阳的东西，但我还是搞不懂你怎么就打破了阴阳平衡。

Q: 传统中医结合阴阳原理来理解人体主要内脏器官的位置和功能，比如说肝脏为阴，胆囊为阳；心脏为阴，小肠为阳；脾为阴，胃为阳；肺为阴，大肠为阳；肾脏为阴，膀胱为阳。

T: 慢点儿说，你这么会儿功夫讲了这么多，给我举个简单的例子吧。我不明白，我只是想知道你是怎么保持阴阳平衡的。

Q: 好吧，阴阳有一些基本的原则，它们又可进一步分为：大阴、小阴、大阳、小阳。它们之间彼此消长、相辅相成，而且可以彼此转化，阳里有阴，阴里有阳。

T: 这么说还是有点儿复杂。不过我还是有点儿懂你的意思了。你现在就你自己的情况为例，给我讲讲阴阳理论吧。

Q: 好吧，这样你会比较好理解。比如说，我没有睡好觉，所以，我阴虚，阳旺。酒是阳，所以我的肝脏里太多阳。吃太多肉和海鲜，这就增加了我的胆囊的负荷，这又是阳气太旺。

T: 这听起来蛮有道理的。可你又怎么解释你过旺的阳气和你身体疼痛有什么关系呢？

Q: 中医把人体看作是一个布满经脉的网络，它连接和平衡各个器官。经脉的一个功能把内脏器官和人的体表相连接。所以，内脏有火，牙就会疼。

Questions 3

1. Do you try to balance your food intake according to the principles above?
2. If you have a physical pain is one of your internal organs to blame? How would you remedy it?

29 房子

Houses

Background Information

The annual growth of property prices in large Chinese cities is 5.3% (Beijing 9% this first quarter alone and 42% in the last three years). According to a survey by Beijing Consumers Association 80% of homeowners are unhappy with their homes and 41.2% complain that their homes have defective designs.

中国大城市的房地产价格年均增长率为 5.3% （北京仅在今年第一季度就增长了 9%，在过去的 3 年里增长了 42%）。北京消费者协会的一项调查显示，80% 的购房者对自己的房屋并不满意，41.2% 的购房者抱怨自己的房子存有设计缺陷。

背景信息

Dialogue 1

Rose: Can you tell me another difference between China and England?

Nick: I suppose it's housing. Certainly in the cities here most people live in flats but we would live in houses.

R: What kind of house?

N: I suppose a typical house would be semi-detached. That's a two-storey house divided into two.

R: We don't have as much land available for housing as you do and we have a much larger population too!

N: They're also different on the inside too! For example, we're more likely to have **wall-to-wall carpeting**[1] whereas you don't.

R: No, we would prefer a wooden or tiled floor.

N: And another difference is that you remove your shoes when you enter a home. We would keep our shoes on and wipe them first on a doormat.

R: How strange you English are!

N: It's your doors that are strange to us! You have two doors and the outermost door is often like a metal cage!

R: What's strange about that? It's just to deter thieves.

N: In Britain the only people who have doors like that are drug dealers! And they're meant to keep the police out not thieves!

R: So what's normal to us is strange to you and what's normal to you is strange to us!

N: It's only strange until you get used to it.

习惯用语 1

① wall-to-wall carpeting：整个室内铺上地毯

R: 能再告诉我一些中国与英国的其他区别吗？

N: 我想是房子。在中国，大多数城市人都住在公寓楼房里，但是在英国，我们则是住房子。

R: 什么样的房子？

N: 典型的房子就是半独立式住宅，分为上下两层。

R: 我们不像你们那样有足够的土地来盖房子，我们的人口数量太多了！

N: 房子内部的格局也不一样。比如，我们喜欢整屋都铺上地毯，可你们却不这样。

R: 我们不铺地毯，我们还是更喜欢木地板或者地板砖。

N: 还有一点不同是，你们进家门要换鞋，而我们则不，只是在进门前在擦脚垫上先擦一下。

R: 你们英国人也太奇怪了！

N: 你们的门才是奇怪！有两个门，外面的那个就像是一个铁盖子。

R: 那有什么奇怪的？不过是为了防小偷罢了。

N: 在英国，只有毒品贩子才会安装那样的两道门。为的是防警察而不是小偷。

R: 所以，对我们来说是很正常的事情，对你们来说却很奇怪。反之亦然。

N: 要是习惯了，就不觉得奇怪了。

Questions 1

1. Would you prefer to live in a flat or house? Why?
2. What else would be strange about the place where you live to a Westerner?

Dialogue 2

Rose: Any other differences between Western and Chinese houses?

Nick: I think our kitchens are bigger than yours. We sometimes have enough space to eat our meals there.

R: I agree our kitchens are very small. Sometimes there **isn't**

enough room to swing a cat^①!

N: Maybe our kitchens are big because we have a lot of **white goods**^② to put in like an oven, dishwasher, fridge and washing machine.

R: That is a lot!

N: Our bathrooms are bigger too. Most would have a bath, shower, wash basin and toilet.

R: We would love to have bigger kitchens and bathrooms but it would make apartments too expensive for us to buy.

N: One thing I've noticed is that we would have a shower before going to work in the morning while you tend to have showers at night.

R: Yes, we think that having a shower last thing at night helps us to sleep. What other differences are there?

N: I know that most Chinese homes would have air conditioning but you would never find that in British homes.

R: That's because your summers are milder than ours and also because our summers are humid.

N: That's right. Our heat is dry so we don't need them. Most people will open the window if it gets hot, which isn't often!

习惯用语2

① isn't enough room to swing a cat：空间非常小
② white goods：白色家电

R: 西方与中国的房子还有其他什么不同吗？

N: 我们的厨房比你们的要大，即使在里面吃饭都可以。

R: 没错，中国的厨房空间的确是太小了。

N: 我们的厨房面积大，也许是因为我们需要放很多家电，比如烤箱、洗碗机、冰箱、洗衣机。

R: 那么多！

N: 我们的浴室也很大。大多数都有浴缸、淋浴、盥洗池、马桶。

R: 我们也想要大的厨房和浴室，可那样的话，房子的总价就更高了，我们就买不起了！

N: 我还发现一件事不同，我们通常都会在早晨上班前洗个澡，可你们则是在晚上洗。

R: 没错，我们觉得，睡前洗个澡，有助于睡眠。还有什么不同吗？

N: 还有就是我发现很多中国家庭都安装空调，可是在英国的家庭你却很难找到。

R: 那是因为你们那里的夏季比较温暖，而且不像我们这么潮热。

N: 没错。我们的那种热是干燥的，所以我们不需要空调。要是人们觉得热了就打开窗户，而我们那里不会经常那么热。

Questions 2

1. Is a bigger bathroom or kitchen a priority for you?
2. Would you like your summers to be dry like Britain's?

Dialogue 3

Rose: I've often noticed that British houses have sloping roofs. Why is that?

Nick: I think it's because of snow in winter. The angle of the roof forces the snow to slide off so that it does not accumulate and put too much weight on the roof.

R: Most parts of China do not get any or little snow so most of our roofs are flat.

N: That's a pity because I really like your traditional roofs with the curling eaves. But they don't seem to be building them like that anymore. In Beijing they've demolished thousands of the old-type buildings.

R: That's because they've got no **indoor plumbing**[1] and most people want to live in modern apartments.

N: It's a shame because I think they look very nice. In England we would restore them.

R: But it's expensive to do that and anyway by building apartment blocks we can house more people.

N: That's progress I suppose but an Englishman's home is his castle and because **we can buy them outright**② we tend to look after them.

R: Do all British houses have gardens?

N: We like gardens so most houses would have one.

R: Is this for growing vegetables?

N: No. We usually like a lawn and some flowers and there is often a hedge or fence that borders it.

R: Do you often spend a lot of time in the garden?

N: Mainly at weekends. Sometimes we might hold a barbecue and invite friends round.

习惯用语 3

① indoor plumbing：室内卫生设备
② we can buy them outright：付全款购买

R: 我发现英国房子的屋顶都是斜坡形的，这是为什么？

N: 我想可能是因为冬天下雪。屋顶的坡度可以使雪滑落下来，而不至于堆积在屋顶，使其负重过大。

R: 中国大部分地区的降雪量几乎为零，或者很少。所以，我们的屋顶大多是平的。

N: 真遗憾！我特喜欢中国那种传统的卷曲的屋檐。可是现在的建筑似乎不再那样了。在北京，数以千计的老建筑都遭到了破坏。

R: 因为那样的房子没有室内卫生间，现在的人们都想住在现代化的公寓里。

N: 那太遗憾了，我觉得它们真的很好看。要是在英国，我们会将其恢复原貌的。

R: 但那样会花费大量的金钱，再说，盖公寓楼，可以住进去更多的人。

N: 这大概是一种进步。不过对于英国人来说，房子就是他的城堡，因为我们是全部买断整个房屋的，所以当然就要好好打理了。

R: 英国的房子都有花园吗？

N: 大多数都有，因为我们喜欢花园。

R: 花园是用来种蔬菜的吗？

N: 不是。通常就是铺些草坪，种点儿花，用篱笆和小栅栏围起来。

R: 你经常花很多时间打理花园吗？

N: 主要是周末打理一下。有时候我们会邀请朋友过来，在花园里一起烧烤。

Questions 3

1. Would you like to live in an old-style Chinese building? Why? Why not?
2. Would you like to have a garden? Why? Why not?

30 盥洗室
Toilets

Background Information

There are many different names for toilets including
lavatory
W. C. (water closet)
bathroom
restroom
john (American English)
ladies
gents
little boy's room

盥洗室的说法有很多种，比如：
lavatory 卫生间
W. C. (water closet) 厕所
bathroom 浴室
restroom 洗手间
john (美式英语) 厕所
ladies 女卫生间
gents 男卫生间
little boy's room 一号

背景信息

Dialogue 1

Nick: I'm glad that my apartment has a Western-style toilet. I just can't get used to a Chinese **squatty potty**[①]!

Laura: What's your problem?

N: I just can't squat! No matter how hard I try I can never get my heels and soles to stay flat on the floor when I squat. My heels just won't touch the ground.

L: We have learnt to squat from childhood so it's natural for us. You Westerners are just not as supple as we are.

N: I wish I could do it but I just can't. I remember going into a toilet once and all the cubicles were only waist high. I thought they were empty until someone suddenly jumped up. It frightened the life out of me!

L: I bet he was frightened too!

N: You know, there is one thing that puzzles me about Chinese toilets, especially those on campus.

L: What is it?

N: Why are the windows made of clear glass so that people from the outside can look in? And why are the doors into toilets always left open?

L: Maybe we just don't see it as a private function.

N: In the U. K. we certainly would! And you would always find toilet paper in our public toilets.

L: If we provided toilet paper in our toilets it would soon disappear!

习惯用语 1

① squatty potty: 蹲坑，中式蹲厕

N: 真庆幸我的公寓里有西式的马桶，我真的无法适应中式的蹲厕。

L: 怎么了？

N: 我蹲不下去啊！无论如何，我蹲下的时候，脚后跟和脚底都无法平稳着地，我的脚后跟碰不到地面。

L: 我们从小就学着蹲着上厕所，已经习惯了。西方人不像我们这样有柔韧性。

N: 我倒想蹲下呢，可是不行啊。我记得有一次去厕所，所有的隔间都是与腰齐高的，我还以为里面没有人，结果突然就站起来了一个。简直吓死我了！

L: 我打赌那个人也吓坏了！

N: 还有一个关于中国厕所的事儿特别让我疑惑，尤其是学校的厕所。

L: 是什么？

N: 厕所的玻璃为什么都是透明的？人们从外面都能看到里面。还有，厕所的门为什么总是开着的？

L: 大概是因为我们并没有把其视为私密之处吧。

N: 在英国，我们可是很在意的！在我们的公共厕所里也可以找到厕纸。

L: 要是我们在厕所里提供厕纸的话，很快就会没有了。

Questions 1

1. Why do you think campus toilets have see-through windows?
2. What can be done to improve the toilets on campus?

Dialogue 2

Tori: Look, I really **need to go**①. Does this place have a bath-room?

Lucy: Of course! Almost all restaurants have bathrooms. I'll show you the way, follow me.

T: OK, thanks. Erm … what is this?

L: The bathroom. Why? Any problem?

T: It's just, I don't know, different. I mean, why can't you sit down? When I go to the toilet I like to sit down.

L: Here we usually crouch down when we go to the toilet, we don't like to sit. Who knows who has been sitting there before you?

T: Yes, I guess I never thought about that. But it still doesn't really explain the footprints on the toilet seat of that department store earlier.

L: In that department store, they were Western style, sit down toilets. Some Chinese don't like them, so they stand on them.

T: You mean they actually stand on the toilet seat? That's why the footprints were there?

L: Yes. They stand, especially if there are no toilet seat covers. This kind of hygiene thing is very important to us. Who knows what diseases can be transmitted through sitting down? We are just trying to be careful.

T: OK, I got it. So, how do I **tackle**② this thing? Should I face the door or face the back wall? Should I stand on the floor or on the porcelain?

L: Usually, we stand on the porcelain and face the door.

T: OK, well this should be good target practice for me. If I didn't know how to aim before, I'm going to have to learn now.

习惯用语 2

① need to go：去厕所   ② tackle：修复，处理

T: 我得去厕所！这儿有吗？

L: 当然了！几乎所有的餐馆都有。我带你去。

T: 好的，谢谢！嗯……这是什么？

L: 厕所啊。怎么了，有什么问题吗？

T: 就是，我也不太清楚，有点儿不一样啊。我的意思是，怎么没有

坐便器？我喜欢用有坐便器的厕所。

L: 在中国，我们上厕所的时候都是蹲着的，我们不喜欢坐着。谁知道马桶之前被谁坐过啊。

T: 也是，我从没想过这个问题。但这也无法解释之前的那个商场里马桶座圈上的脚印啊。

L: 那个商场安装的是西式的坐便器。有些中国人不喜欢那样，所以就站在上面了。

T: 你是说他们蹲在了马桶座圈上，所以才会留下脚印？

L: 没错。尤其是在没有马桶座圈的情况下，更会如此。个人卫生还是很重要的。要是坐在上面，谁知道会染上什么疾病啊？我们只是尽量小心点儿罢了。

T: 好吧，我明白了。那我现在该怎么办？面冲墙还是门？我应该踩在地上还是这个磁砖上？

L: 通常，我们踩在白瓷部位，面朝门。

T: 好吧，这可是个练习瞄准的好机会。要是之前不知道怎么瞄准，现在必须得学了。

Questions 2

1. Do you prefer Chinese-style toilets or Western-style toilets? Why?
2. Have you any amusing stories about toilets to tell?

Dialogue 3

Lucy: Remember, I mentioned before, to keep tissues in your bag?

Tori: Yes, I remember. Of course, nowadays I always keep tissues in my bag, just in case.

L: Well, this park may be the perfect opportunity to use it. I've hardly been to any parks where they have toilet paper in the bathrooms.

T: Yeah, I've noticed that. Why?

L: I'm not sure. Maybe they are trying to save money. You know, the entry fee is very low. They need to get the money back somehow.

T: So, do all Chinese carry tissues with them everywhere? Because if they need the toilet when they are out of the house, they need to be prepared?

L: Yes, I guess so. I've seen on American TV shows, sometimes the ladies go to the toilet and don't take any paper.

T: Yes, usually Western toilets are **fully stocked**[①] with toilet tissue. So, we don't really worry about that.

L: But, when they look in the dispenser, there is no tissue, so they ask the person in the next stall.

T: Yes! This is usual, actually. If it happens that the person next door is awful, maybe they won't spare you a piece. Now that is really embarrassing!

L: Well, at least that can never happen to you in China. Now because you listened to me, you are always prepared.

T: Yes, thanks. It's really good advice. I think everyone who comes here should be prepared. If you aren't, it's a nasty shock!

习惯用语 3

① fully stock：准备充足

L: 你记不记得我原来说过要随身携带纸巾?

T: 记得。我现在就是随身携带，以防万一。

L: 好极了，这回在这个公园可算有用武之地了。我就没看到过哪个公园在洗手间里提供厕纸的。

T: 是啊，我也注意到了。为什么啊?

L: 我也不知道。可能是为了省钱吧。门票这么便宜，人家总得赚钱啊。

T: 那所有的中国人都是随身携带纸巾吗? 要是他们外出的时候想去

厕所，总得准备些吧？

L: 我想是的。我在美国的电视节目上看到过，有时候女孩子想去厕所，却没带纸。

T: 没错。通常西方的厕所都准备有手纸的，所以我们也不用担心。

L: 但当他们看抽纸箱里没有手纸的时候，那就只能向旁边的人借了。

T: 没错，这很常见。不过，要是你旁边的人不怎么样，那他们可不会把纸借给你。那才叫尴尬呢。

L: 没错，不过至少在中国，这不会发生在你身上。因为你听了我的建议，随身准备。

T: 是的，多谢。还真是个不错的建议。我觉得所有来中国的人都要随身携带。要是不带的话，那就不好办了。

Questions 3

1. What would you do if you went to the toilet and found you had no toilet paper?
2. What would you do if there was no public toilet nearby?

31 家庭

Families

Background Information

The immediate family consists of yourself, any brothers and sisters and your parents. The extended family includes uncles, aunties, cousins and grandparents. In 1985 Chinese household size was 4. 5 people per house. In 2000 it was 3. 5 and is projected to decrease further to 2. 7 by the year 2015. This means that today China has 80 million more households than it would otherwise have had.

The household size decrease results from social change: especially, population aging, fewer children per couple, an increase in previously non-existent divorce, and a decline in the former custom of multi-generation households.

核心家庭的成员主要由你自己、兄弟姐妹、还有你的父母组成。大家庭则还包括叔叔、阿姨、表兄妹、祖父母。在 1985 年，中国的家庭大小一般为每户 4.5 人。在 2000 年，这一数字为 3.5 人，而到 2015 年预计会进一步减少到 2.7 人。这就意味着，现在中国的住户要比过去多出 8 千万户。

家庭成员人数的减少是社会演变的结果：尤其是人口老龄化、每对夫妇所生孩子的减少、增加原先根本不存在的离婚，以前几代同堂的家庭越来越少。

背景信息

Dialogue 1

Michael: So, tell me about your family. You're from Beijing, right?

Sally: Yes. I'm a Beijinger. I guess my family are pretty traditional.

M: How so? You still live with your parents, don't you?

S: Yes, I do, but that's usual. I mean, I'm not married yet, and my folks live in the same town, so why wouldn't I?

M: Well, it's different overseas, we can't wait to get out and live alone or share a house with friends.

S: I've noticed that all of you guys live together, or John, who lives alone. I just thought that's because you are away from your home country, you try to get together. Maybe it makes you think of your family.

M: No, that's not right. We can't wait to **flee the nest**①, to get out on our own. It gives us a sense of independence.

S: But unless you had to, why would you leave your folks? I don't understand, I mean, they look after you, care for you, do your washing, and so on.

M: Yes, maybe I can see that. But overseas, it's usual for someone to move out, when we go to university, and never return.

S: So when you hit 18 you just leave them? They are your parents!

M: I know, but **it's just the norm**②. I mean I did move back for 3 months, until I got a job, but as soon as I was earning I got a place of my own.

S: I guess I must be lucky. I could stay at home during university and didn't have to live on campus, so I've never had the experience of sharing a house with others. To be honest, I think staying at home is good. At least until you get married.

习惯用语 1

① flee the nest：空巢，离开家
② it's just the norm：惯例，与往常一样

Everyday Life 日常生活

M: 跟我说说你的家庭吧。你来自北京，对吗？

S: 没错，我是北京人。我想，我的家庭是非常传统的那种。

M: 怎么这么说？你还跟父母住在一起，是吧？

S: 是的，这很常见。我是说，我还没有结婚，而我的父母都住在同一个城市，所以我为什么不和他们一起住呢？

M: 在国外可不是这样。我们都迫不及待地想要搬出去单住，或者与朋友合住。

S: 我注意到了，你们都住在一起，只有约翰单独住。我还以为你们是因为离家在外所以想聚在一起住呢，也许这样会让你们更能体会家的感觉。

M: 不是，我们只是想赶紧搬出来，靠自己生活。这让我们有一种独立自主的感觉。

S: 如果不是迫不得已，那你为什么要离开父母呢？我是说，我不明白，因为住在一起，他们可以照顾你，关心你，帮你洗衣服。

M: 是的，我知道。但在国外，孩子上了大学以后就会离开家，再也不回去了。

S: 所以，等你到了 18 岁你就会离开他们？他们可是你的父母亲啊！

M: 我知道，这没什么可奇怪的。我是说，我每三个月还会回去一次，直到我找到工作，一旦我开始自己挣钱了，我就会拥有一个属于自己的家。

S: 我想我一定算是幸运了。大学期间可以待在家里，而不必住在学校。所以我从来没跟别人合住过。老实说，我觉得住在家里很不错。至少要一直住到结婚的时候。

Questions 1

1. Do you like living at home? Why? Why not?
2. Would you like to share a flat or live alone?

Dialogue 2

Nick: I suppose most families would get together at Christmas.

Kate: That's your equivalent of our Spring Festival. So tell me what you do then?

N: The main event is Christmas dinner. We call it dinner but it's held at lunchtime.

K: So what do you eat?

N: The main item is turkey and usually the bigger the better! It would be put in the oven early on Christmas Day as it takes several hours to cook.

K: We never eat turkey in China. I know Americans eat it at Thanksgiving but why do you?

N: It used to be a goose that was traditionally eaten as you may recall from Dickens's *Christmas Carol* but I think turkeys **caught on**[①] because they are bigger and could cater for the whole family. We would stuff them with breadcrumbs and seasonings and then would come all the trimmings.

K: The trimmings! What's that?

N: All the food that goes with the turkey such as roast potatoes, Brussel sprouts, peas, carrots, cranberry jelly and then there would be Christmas pudding and custard to follow.

K: That's a lot of food to eat!

N: And not forgetting mince pies! Then at 3 pm we would relax in front of the TV and listen to the Queen's Speech.

K: Next Christmas be sure to invite me, OK?

N: Sure.

习惯用语 2

① catch on：流行

N: 我想大多数家庭都会在圣诞节时聚在一起。

K: 就跟我们过春节时一样。跟我说说你们圣诞节都做些什么?

N: 最主要的就是圣诞晚宴。尽管我们称其为晚餐,其实它从中午的时候就开始了。

K: 你们都吃些什么?

N: 主菜是火鸡,而且是越大越好!通常一大早就会把它放进烤箱里烹制,因为这要花费好几个小时。

K: 我们在中国从没吃过火鸡。我知道美国人在感恩节的时候吃火鸡。但你们为什么在圣诞节的时候吃呢?

N: 传统的吃法其实是烧鹅,这会让你回忆起狄更斯的《圣诞颂歌》。但现在更流行吃火鸡,我觉得是因为它的个头大,够全家人一起吃。我们会把面包屑和各种调料塞进它的身体里,然后和各种配菜一起吃。

K: 各种配菜?都是些什么?

N: 与火鸡相搭配的各种食物,比如烤土豆、抱子甘蓝、豌豆、胡萝卜、越橘果冻,随后还有圣诞布丁和奶油蛋糕。

K: 那么多好吃的啊!

N: 还有别忘了肉馅饼。下午 3 点的时候,我们会围坐在电视机前,休息一下,听女王演说。

K: 下次过圣诞节的时候一定要邀请我啊!好吗?

N: 没问题!

Questions 2

1. Do you like family meals? Why? Why not?
2. Would you like to eat a British Christmas meal? If so, where do you think you could get it?

Dialogue 3

Nick: So, you live in a big apartment, right?

Sabine: It's OK. About 150 square metres, we're lucky I guess.

N: But that's so huge! My apartment is much smaller, so why are you so upset?

S: Oh, I'm not upset, but you don't understand. Sometimes I wish I just had time for myself. I don't just live there with my husband.

N: But, it's your house! Who do you live with? You mean you share with your friends?

S: No, I live with my husband's parents. They like to be close to us, we help them and they help us.

N: Oh! In England that would be so weird! After you marry, maybe even before, we move away. We need to have our own space.

S: So, you don't live with your parents after marriage? I think that is strange, why not have them close? They are useful and they can look after you.

N: But, I don't understand, you are married. Why do you need someone else to look after you?

S: It's good to have a happy relationship, you know how hard I work. I need someone to cook dinner for me and my husband. If my mother-in-law wasn't there, I'd have to hire an **ayi**[1] to do all of the housework.

N: So you mean you treat your mother-in-law like a slave?

S: No, no. I don't mean that. I mean that she wants to help, we are all so busy, she has nothing to do, so she enjoys living with us. Here in China, that's just what we do.

习惯用语 3

① ayi：阿姨，保姆

N: 你住在一个大公寓里，对吧？
S: 还可以吧，大概 150 平米。我觉得我们很幸运。
N: 那么大！我的房子可小多了，那你还有什么不高兴的？

S: 不是不高兴，你不理解。有时候我真希望能有自己的时间空间。你知道，和我一起住的不只有我的丈夫。

N: 但那是你的房子啊！还有谁跟你一起住？你是说跟朋友合住？

S: 不是。还有我丈夫的父母。他们想跟我们住在一起，彼此好有个照应。

N: 天哪！这要是在英国简直太荒唐了！在结婚后，甚至婚前，我们就搬出去住了。我们需要自己的空间。

S: 你婚后并不是和父母住在一起？我倒是觉得这很奇怪，为什么不让他们在身边呢？他们可以照顾你啊。

N: 我不明白，你已经结婚了，为什么还要别人来照顾你呢？

S: 家庭成员之间关系融洽很重要。我工作很辛苦，我需要有个人来帮我和丈夫做饭。要是我婆婆不在的话，我就得雇一个阿姨来做这些家务。

N: 你是说你像对待奴隶一样对待你的婆婆？

S: 不，我不是那个意思。我是说，我们工作很忙，她又无事可做，很想来帮助我们，她喜欢和我们住在一起。在中国，就是这个样子的。

Questions 3

1. When you get married would you have your in-laws living with you? Why? Why not?
2. Would you employ an ayi? Why? Why not?

W ork

32 工作

Background Information

England has an average working week of 48 hours while China has 40. According to www.zhaopin.com when the boss is away Chinese office workers spend their time by

- Online surfing 80%
- Online chatting 58%
- Blogging 19%
- Playing games 15%

英国平均每周的工作时间为 48 小时，而中国则为 40 小时。智联招聘（www. zhaopin. com）的调查显示，当老板不在的时候，中国员工会把时间花在：

- 浏览网页，占 80% 的时间
- 网上聊天，占 58%
- 写博客，占 19%
- 玩游戏，占 15%

背景信息

Dialogue 1

Diana: You do like your boss, don't you?

Mark: Oh, don't even get me started about my boss!

D: Why not? Don't you get along with him?

M: No one gets along with him! Even his wife left him because she couldn't take it anymore. It's not just that, my job involves a lot of time at the computer, many hours, it can be very tiring. We're all thinking of **going on strike**①, he expects too much of us!

D: Go on strike? It's illegal here. We try to never protest about such things. Why not quit?

M: Never! We don't just quit if we hate our bosses. We try to talk to them to understand what the problem is between us. And also, it was my childhood dream to work on cartoons.

D: I'm confused. Why do you continue with a job that gives you so much pressure and with a boss you hate? Here we don't talk to our bosses or go on strike, we just simply quit and try to find something else.

M: Maybe you are right. I like the money and the title, and the idea of working on cartoons, but this job is giving me an ulcer. If I keep working like this I'm going to be dead before I'm 40!

D: Then, if it's your usual way of doing things, I think you should have a chat with your boss, tell him how you feel.

M: Maybe you are right. I think I'll do that. See, quitting is not always the only option. Maybe if I talk to him he will see **the error of his ways**②.

习惯用语 1

① go on strike：罢工，拒绝工作
② the error of sb.'s ways：某人犯的错误

D: 你很喜欢你的老板，对吧？

M: 哦！别跟我提我的老板！

D: 为什么？你跟他相处得不好？

M: 没人能跟他合得来。就连他的妻子也终因无法忍受而离开他了。不仅如此，我的工作需要长时间对着电脑，很累。我们都在考虑罢工呢，他对我们要求太高了。

D: 罢工？这是违法的啊。我们不会因为这个而抗议。干嘛不辞职啊？

M: 决不。我们不能因为讨厌老板就辞职。我们会找他们谈，告诉他们我们之间的问题出在哪里。另外，在动漫领域工作是我儿时的梦想。

D: 我糊涂了。这份工作压力那么大，你的老板又那么惹人讨厌，你为什么还要继续工作呢？换作是我们，可不会去找老板谈，或者是罢工，我们只是辞职然后再找一份新的工作罢了。

M: 也许你是对的。我喜欢钱、头衔以及在动漫领域工作的感觉。但这份工作让我痛苦。要是这样继续工作下去的话，我肯定在 40 岁前就死掉了。

D: 要是像你通常的做事风格，我想你应该跟你的老板谈谈，告诉他你的感受。

M: 也许你是对的。我会这样做的。要知道，辞职并不总是唯一的办法。要是我和他谈谈，也许他能够发现自身的错误之处。

Questions 1

1. What do you do if you don't like your boss?
2. How do you deal with work problems?

Dialogue 2

Ian: Hello sir, nice to see you. You don't look too happy, if you don't mind me saying so. What's the problem?

John: Ian, I've told you before, you don't need to be so formal. Please call me John. Anyway, we've got a bit of a crisis in our

Beijing office at the moment.

I: Oh! I thought the company was doing very well. Is business bad?

J: No, no, nothing like that. There won't be any **layoffs**[①], we are doing fine, just some staff problems. Do you remember Martin Price?

I: Yes, I do. He only went out there a few months ago. You transferred him over here from Chicago, didn't you?

J: Well, it seems that he is **not cut out for the job**[②]. We really need a local to run the Beijing office. I wish I had known that before I took him on, but he seemed well qualified, even though he didn't speak Chinese. It turns out he has no idea about doing business in China. When I called and relieved him of his position, he didn't take it very well.

I: Yes, that is a problem here. For a foreigner to succeed over here they need to speak at least some Chinese and understand the way we Chinese do business. It's totally different from Western countries. What did he say?

J: He asked me why he was getting sacked. I told him that his performance wasn't what we had expected, and that we were not going to sack him but re-assign him to another office. To be honest, his biggest problem was the wining and dining of clients. In America, we try to do business quickly, during maybe a **power lunch**[③], in 15 or 20 minutes. Here, there are so many long dinners, and so much drinking!

I: Yes, I understand. That's just the way we do it, like I said, if you are going to do business here, you must understand. And the drinking is a very important part of it, if you refuse to drink with your client, he may think you have no respect for him. Then, you lose the deal. How much longer will he be out there?

J: Yes, I know. It's seems Martin isn't very good at schmoozing. At the moment no one is there. He said he would **hand in his notice**④. So I asked him why he wanted to resign. First, he told me that we didn't pay him enough and that he had found a better job in another company. Then he said, and I quote, "You can take this job and **shove it**⑤! I quit!"

I: Oh? How can I help?

J: Well. There is an important empty desk in Beijing. I'd like you to sit behind it.

习惯用语 2

① layoff：富余人员，临时解雇
② not cut out for the job：不能胜任工作
③ power lunch：边吃午餐边讨论工作
④ hand in sb.'s notice：提前通知（离职前要提前一周或一个月通知）
⑤ shove it：离职，洗手不干（表示轻蔑）

I: 你好先生，很高兴见到你。要是您不介意的话，我想说，您今天看上去不太高兴啊。怎么了？

J: Ian，我之前就跟你说过了，不用这么见外。直接叫我 John 就可以了。现在，我们的北京分公司出现了一些运营危机。

I: 啊？我还以为公司一直运转的不错呢。生意不好了？

J: 不，不是。没人被裁，我们运转得很好，只是有些员工纠纷。你还记得 Martin Price 吗？

I: 记得。他不就是几个月前刚到北京分公司的嘛，当时你把他从芝加哥调到这里的，对吧？

J: 嗯，他似乎有点儿不能胜任这份工作。我们需要一个北京当地人来负责北京办公室的运营。要是我能提早了解他就好了，可他当时确实很合格，尽管他不会说中文。可事实是，他似乎并不懂得如何在中国做生意。我给他打了电话，把他从那个职位撤下来，他好像不太高兴。

225

I: 嗯，那可真是个问题。对于一个外国人来说，要想在这里做生意，至少他得会说一点儿中文，还要懂得中国人做生意的方式。这与西方国家完全不同。他当时说什么？

J: 他当时问我为什么把他解雇。我说是由于他的工作表现并不如人所愿，并且我们并不是要解雇他，而是要给他重新安排职位。老实说，他最大的问题就是跟客户吃吃喝喝。在美国，我们会抓紧时间做生意，即使是午饭的15、20分钟也不放过。可在这里，总是在吃吃喝喝。

I: 嗯，我明白。这就是我们做生意的方式，就像我所说的，要想在这里做生意，你就得理解。吃喝就是很重要的组成部分，要是你不请客户吃饭，他们会觉得你不尊重他。那你就会失去这个买卖了。他还会在那儿待多久？

J: 我知道了。看来 Martin 不太擅长公关啊。现在没人在那里。他说他会递交辞呈的。我问他为什么辞职，首先，他跟我说，他的工资太低，他已经在别的公司找到新工作了。随后他说，他原话是这样的，"你来干干看，我是不干了！我辞职！"

I: 噢？我能帮上什么忙吗？

J: 嗯，北京有个很重要的职位空缺，我希望你去填补。

Questions 2

1. What do you think of drinking with clients?
2. Would you like to work for a multinational company? Why? Why not?

Dialogue 3

Polly: Oh! Go home and change. I'll **cover for you**[①], but do it quickly before the boss comes in. Whatever made you think to dress like that for work?

Nathan: Polly, what day is it?

P: Now I know you have really gone mad, you don't even know what day it is. It's Friday, silly.

N: Exactly! Didn't you know, the office has just introduced a new policy to make the foreigners working here feel more at home? It's called "Casual Friday"?

P: Casual Friday? I've never heard of that. What is it?

N: Well, because we must wear suits every day of the week, the boss thought it would be nice to give us one day off from it, hence Casual Friday. You can wear whatever you want.

P: So that's why you're wearing jeans and trainers, I get it now. So, is it common in England?

N: I think the idea started originally in America, but some companies in England do it too. For example, in my old company, on Fridays you could wear what ever you wanted, but you had to pay to do so.

P: Pay? Why and how much?

N: Not much, just a couple of pounds, then the office would put all the money together and use it for something good. Sometimes for our staff Christmas party or sometimes we would give it to charity.

P: That sounds like a good idea, maybe we could try that here. If we charged 10 yuan per person, there's 10 of us here, so that's 100 yuan.

N: Great idea! We could put the money towards our Spring Festival Party and go to a really amazing restaurant next year. I'll go and **clear it**② with the boss.

> **习惯用语 3**
> ① cover for sb.：暂时代替某人照管
> ② clear sth.：将某事提交审批

P: 天哪！赶紧回家换衣服！我来替你一会儿，快点儿啊，别等老板来了。你怎么穿成这样就来上班了？

N: Polly，今天星期几？

P: 我现在知道你算是彻底疯掉了。你竟然不知道今天星期几？星期五啊！傻东西！

N: 对啊！你不知道嘛，为了让外国人在这里工作感觉更像是在家里，公司刚刚颁布的新政策，叫做"休闲星期五"。

P: 休闲星期五？我没听过啊。什么意思？

N: 因为我们平时工作必须穿正装，老板觉得要是能给我们解放一天就好了，所以就提出了"休闲星期五"。你想穿什么都行。

P: 我明白了，所以你才穿了仔裤和运动鞋。这在英国很常见吗？

N: 这个想法起源于美国，但一些英国公司也这么做。比如我原来的公司，周五就可以随便穿，不过得交钱。

P: 交钱？为什么？交多少？

N: 不多，就几英镑。公司把这些钱收集起来做些有意义的事。有时候用作员工的圣诞晚会，有时候捐给慈善团体。

P: 这听起来倒是个好主意，或许我们也可以试试。每人收 10 块钱，10 个人就是 100 块钱。

N: 好主意！我们可以把钱用在春节派对上，明年就可以去一个好餐馆了。我这就去跟老板申请。

Questions 3

1. What kind of clothes do you have to wear in the office?
2. Should you be allowed to wear casual dress? Why? Why not?

33 运动

Sports

Background Information

According to *China Youth Daily* each person in China has only 0.67 square metres of sports grounds, a quarter of that in the West. A 2005 report by the Beijing Statistics Bureau found that people in Beijing spent only 30 minutes doing sport at the weekend compared with 5 hours spent watching TV.

据《中国青年报》报道，中国人均只有 0.67 平方米的运动场地，是西方国家的四分之一。北京统计局一份 2005 年的报告指出，北京人周末仅运动 30 分钟，而要看 5 个小时的电视。

背景信息

Dialogue 1

Janey: Football is England's national sport, isn't it?

Nick: Yeah, we invented it. Rumour has it that Anglo-Saxons used the heads of captured Danes as footballs.

J: But what did they use as footballs when they ran out of heads?

N: Inflated pig bladders would you believe!

J: So football has always been popular in England, right?

N: Not quite. It was actually banned in 1477 because it interfered with young men learning to be archers.

J: Wow. I suppose national defence had to come first then. What about the modern game? How did that start?

N: It wasn't until 1863 that football **became as we know it now**①.

J: Why? What happened then?

N: The rules of the game were formulated by a group of public school men and that led to the formation of football clubs and leagues.

J: The Premiership is supposed to be the best league in the world but England haven't won the World Cup since 1966. Why is that?

N: I think there are two reasons for that. First, there are a lot of foreigners who play in England which means less room for English players and secondly, we've never had a good manager.

J: I hope that will change in the future. Who do you think will win the Premiership?

N: Last year it was **a two horse race**② between Manchester United and Chelsea.

J: It would be nice if a different team won this year. By the way, can you tell me what a "**hat trick**③" means?

N: It means a player who scores three goals in a game.

J： 足球是英国的国球，是吧？

N： 是的，我们英国发明的足球。传说，当时盎格鲁-撒克逊人用被抓获的丹麦人的头颅当作球来踢。

J： 但是人头用完之后，他们用什么当足球呢？

N： 你信吗，他们就用充满气的猪的膀胱当球。

J： 所以足球在英国一直很受欢迎，是吗？

N： 并不完全是。实际上 1477 年的时候明令禁止踢足球，因为它干扰了年轻人，使他们不愿去学习成为射手。

J： 哇。我觉得那时候国家防御还是应该排在第一位的。现代比赛怎么样呢？是如何开始的呢？

N： 直到 1863 年，足球才变成我们现在采用的形式。

J： 为什么？那时发生了什么事？

N： 比赛规则是由一群公立大学学生制定的，这就形成了足球俱乐部和联盟。

J： 英超联赛被认为是联盟中最好的，但是英国从 1966 年以后还没赢过世界杯。为什么呢？

N： 我认为这有两个原因。首先，英国有很多外籍球员，这就意味着减少了英国球员的机会。第二，我们从来没有一位优秀的经理人。

J： 我希望将来会有所改善。你认为谁会赢得英超联赛？

N： 去年是曼彻斯特和切尔西之间的角逐。

J： 如果今年会有另一支球队赢就好了。顺便问一下，你能告诉我什么是"帽子戏法"吗？

N： 就是一名球员在一场比赛中获得三个入球。

Questions 1

1. Which is China's national sport?
2. What's your favourite sport? Why?

Dialogue 2

Janey: Someone at the coffee shop yesterday, an Englishman, was talking about something called "Cricket". He said it's a really popular sport in England, but I've never heard of it.

Nick: Ah, yes, cricket! Maybe not as popular as football, but still one of our national sports. I can't believe you've never heard of it. Don't they show "The Ashes" on TV here?

J: The Ashes? It sounds like something is on fire, what is it?

N: You are **barking up the wrong tree**^① there, it's a cricket competition, probably the most famous one in the world. It's played between England and Australia and dates back to 1882.

J: It has been going on for a long time. So how do people play cricket? It's a ball game, right?

N: Yes, right. It's played with a bat, made from special willow wood and a very heavy ball, which is wrapped in red leather. The batsman has to wear lots of body protection in case he is hit by the ball. There are two batsmen who, after one of them has hit the ball, must run between two wickets as many times as they can.

J: I see. If the ball's so heavy I think it could probably break bones. It sounds as dangerous as American football!

N: The pads they wear help to prevent serious damage. Anyway, cricket didn't just start with the Ashes, it became popular in England all the way back in the 16th century. No one knows exactly who started it, but it's believed that it was invented by children living in an area of woodland that stretches across the south-east of England. But some people claim that it came from India and was brought to England via Persia.

J: Hmm … most old things have too many stories of origin! It's interesting if it was invented by children. It must be a fun game to play. Do you have any of the equipment needed to play with you here?

N: Yes, I do. Do you fancy going over to the sports field? It won't be a real cricket field, but we can pretend. It'll still be fun. I can teach you how to "knock it for six!"

J: Knock somebody for six … you've taught that in class, haven't you? You said it means to totally surprise some one. It's really about cricket?

N: Yes. If you knock the ball for six, it goes out of the field and you automatically get six points, or runs as we call them. Kind of like a home run in baseball. It's a shock if someone gets one, because it's so difficult!

习惯用语 2

① bark up the wrong tree：找错（弄错）目标了

J: 昨天咖啡店里有个英国人，在讨论什么"板球"。他说这在英国是非常流行的运动，但是我从没听说过。

N: 啊，是的，板球！可能没有足球那么流行，但也是我们的国球之一。我真不敢相信你从来没听说过。这儿的电视上没有播放过"灰烬杯"吗？

J: "灰烬"？听起来像是什么东西着火了，它是什么？

N: 你理解错了，是一项板球比赛，可能是世上最著名的了。英国和澳大利亚之间从 1882 年就开始有比赛了。

J: 已经持续了很长时间。那怎么玩儿板球呢？这是球类运动吗？

N: 是的。用一个由柳木制成的球棒和一个红色皮革包着的很重的球来玩儿的。击球手戴着许多身体护具以防被球击中。有两个击球手，其中一个击中球后，必须在两个门柱之间尽可能多地来回跑动。

J: 我知道了。如果球这么重，我觉得很有可能把人打骨折了。听起来跟美式橄榄球一样危险！

N: 他们戴着衬垫有助于保护身体，防止严重的伤害。不管怎么样，板球不是从灰烬杯比赛开始的，它在英国变得流行可以追溯到 16

Leisure 休闲

世纪。没人知道谁发明的板球，但可以确定的是，它是由住在英国东南部林地的小孩发明的。但是有些人声称是从印度经波斯传到英国来的。

J: 嗯……大部分古老的东西都有很多起源故事的版本！如果是由小孩发明的那就很有趣。肯定是个有趣的游戏。在这跟你玩儿的话你有什么必需的装备吗？

N: 有啊。你想去运动场玩儿吗？不是真正的板球场地，但我们可以假装它是。我可以教你怎么"knock it for six"！

J: 你在课上教过 Knock somebody for six，不是吗？你说它的意思是让别人彻底惊讶。真是关于板球的吗？

N: 是的。如果你击中球，球飞出场地，你就可以得到 6 分，否则就得跑。有点儿像棒球的全垒打。如果有人得了一个 6 分，就会让人很惊讶，因为这太难了。

Questions 2

1. Have you ever heard of cricket before?
2. Do you think that you would like to play it?

Dialogue 3

Janey: I need to write a paper for the university English magazine about sports. I was hoping to do it on sport in England. Do you think you could help me?

Nick: No problem, there are many popular sports in England, both **home grown**① and imported. What do you want to know?

J: Well, which sports do you think are the most popular?

N: That's a tough one … sports are an important part of English life, even if it's just watching them on TV. Maybe cricket, football, lawn tennis and rugby are the most popular. My personal favourite is rugby.

J: I think rugby is a little difficult for me to write about, not many people in China know about it, so maybe they won't understand my article.

N: Why don't you write about **lawn tennis**②? After all, everyone has heard of Wimbledon. Wimbledon is the oldest major tennis tournament, it began in 1877. Also, there are many traditions that go with Wimbledon.

J: Really? We just know it for the famous tennis players, like Maria Sarapova, lots of guys like her.

N: There is something else lovely about Wimbledon, not just the beautiful women. The strawberries!

J: What do strawberries have to do with tennis?

N: Strawberries and cream, to be precise. It's traditional for visitors who watch live tennis to eat strawberries and cream while they watch.

J: Really? That sounds delicious. Maybe I can change my article a little, make it not just about sport in England, but the traditions that go along with that sport.

N: Great idea. You can talk about tennis and strawberries, beer and football, and cricket and cucumber sandwiches!

习惯用语3

① home grown：本国出产的
② lawn tennis：草地网球（在草地上打网球，与红土或沥青场地相对）

J: 我要为大学英语杂志写一份体育报道。我正希望到英国做这个报道呢。你觉得有什么能帮我的吗？

N: 没问题，英国有很多流行的运动，既有本土的，又有引进的。你想知道什么？

J: 嗯，你认为什么运动最受欢迎？

N: 很难说啊……运动是英国人生活中重要的一部分，即使是在电视上看。可能板球、足球、草地网球和英式橄榄球是最受欢迎的。

Leisure 休闲

我最喜欢的是橄榄球。

J: 我觉得橄榄球对我来说有点儿难写，中国有很多人都不知道橄榄球，所以他们可能不能理解我的文章。

N: 你为什么不写草地网球呢？毕竟，每个人都听说过温布尔登。温布尔登是最早的重要的网球锦标赛，开始于1877。而且伴随着温布尔登比赛，产生了很多惯例。

J: 真的吗？我们只知道著名的网球选手，像 Maria Sarapova，很多人喜欢她。

N: 温布尔登还有很多别的有趣的事，不是只有美女。还有"草莓"!

J: 草莓和网球有什么关系？

N: 草莓冰激凌，确切地说。对来观看现场比赛的观众来说这是一项传统，他们一边看比赛一边吃草莓冰激凌。

J: 真的吗？听起来很美味。也许我可以改变一下我的文章，不仅写英国的运动，也写下与之并存的传统惯例。

N: 好主意。你可以谈论网球和草莓，啤酒和足球，板球和黄瓜三明治!

Questions 3

1. Are there any special foods connected with Chinese sports?
2. Can you name any sports that are home grown in China?

34 假期
Holidays

‖ **Background Information** ‖

England has 20 statutory holidays a year while China has 10.

China has three holiday Golden Weeks in a year when travel is popular（May Day，National Day and Spring Festival）. These three weeks account for 25 percent of the annual domestic travel market. 357 million Golden Week trips were made in 2006. The average amount spent by each traveller in 2006 was 447 yuan.

英国一年有 20 个法定节假日，中国有 10 个。

中国一年有三个适合旅游的假日黄金周（五一、十一和春节）。这三个黄金周占全年国内旅游市场的 25%。2006 年黄金周内旅游达 3.57 亿人次，而人均旅游消费为 447 元。

背景信息

Dialogue 1

Kate: Spring Festival is the most important festival to us. Do you know anything about it?

Nick: Not too much. Just that it's the lunar calendar's New Year. It's different from New Year in the West.

K: Yes, that's right. It's the first day of the lunar month, and it originated in the Shang dynasty at around 1600 B.C. It's based on the people's sacrifices to their gods and ancestors and the end of a year and the beginning of a new one.

N: The Shang dynasty? That's a long time ago. I had no idea it had such a long history. And it's always the first day of the lunar month?

K: **Strictly speaking**①, it starts every year in the early days of the 12ᵗʰ lunar month and will last to the middle of the 1ˢᵗ lunar month of the next year. The most important days are Spring Festival Eve and the first 3 days.

N: I see that you and your family are all at home. Don't you have to work during Spring Festival?

K: No, we don't need to work. The government has made it a 7 day national holiday. We usually spend the time with family and friends.

N: Do you do anything special on the other days?

K: Yes, on the 8ᵗʰ day of the 12ᵗʰ lunar month we eat some very special food. It's called laba congee. We roll balls from the glutinous rice, fill them with berries or beans and boil them in water. Nowadays, you can even buy glutinous rice balls filled with chocolate!

N: I'd rather eat chocolate on its own. Are there any other significant days?

K: The 23ʳᵈ day of the 12ᵗʰ lunar month we call Preliminary Eve.

At this time, people offer sacrifices to the kitchen god. But, to be honest, **in this day and age**②, most families cook wonderful food, get together and enjoy themselves.

N: Well, I'm happy to be here for all of the celebrations. Someone told me we will watch an interesting TV show tonight.

K: Yes, CCTV has a special New Year variety show with all of our **top stars**③. Singers, dancers, comedians, everything!

N: I heard that most young people find it boring as its been the same rigid format for about 30 years. They prefer to do something online.

习惯用语 1

① strictly speaking: 准确地说，严格地说
② in this day and age: 在今天
③ top star: 明星

K：春节是我们最重要的节日，你知道关于春节的情况吗?

N：不是很多，仅仅知道它是阴历的新年。它跟西方的新年不同。

K：是的，没错。春节是农历的第一天，它起始于公元前 1600 年左右的商朝。春节是人们供奉他们的神和祖先，以及一年的终点和新年的开始。

N：商朝? 那是很久以前了。我不知道它有这么长的历史了。那春节总是阴历元月的第一天吗?

K：严格来讲，它开始于每年农历 12 月的上旬，将持续到来年农历元月的中旬。最重要的日子是除夕和开始的 3 天。

N：我知道你和你的家人都在家里。你在春节期间不工作吗?

K：是的，我们不需要上班。政府规定有 7 天国家法定假日。我们通常跟家人和朋友在一起。

N：你们在其他的日子有什么特殊的活动吗?

K：是的，在农历 12 月的初八我们吃很特殊的食物，叫做腊八粥。我们还用糯米滚元宵，里面装上果馅或者豆沙，然后在水里煮。现

在，我们甚至可以买到巧克力馅的。

N: 我更喜欢吃巧克力的。还有什么其他重要的日子吗？

K: 农历 12 月的 23 日我们称作小年。这一天人们给灶王爷上供。但是，坦率地讲，在今天这个时代，大多数家庭会烹调美味佳肴，大家聚在一起自己享用。

N: 我很高兴能在这儿参加所有的庆祝活动。有人告诉我今晚可以看到很有意思的电视节目。

K: 是的，中央电视台将有一个特别的春节晚会，由我们最著名的明星，如歌手、舞蹈演员、喜剧演员等参加演出！

N: 我听说大多数年轻人都觉得它很乏味，因为它 30 年形式都没变。他们更喜欢上网。

Questions 1

1. How do you spend Spring Festival?
2. What do you do instead of watching CCTV?

Dialogue 2

Nick: Which Western holiday or festival do you like the best? I bet it's Christmas, Christmas is very popular in China with the under 30's.

Kate: I would have to say Valentine's Day. It has a really interesting history; 800 years before Valentine's Day was set up the Romans practiced a pagan celebration in mid-February. Part of the celebration was a lottery where young men would draw the names of teenage girls from a box. The girl assigned to each young man in that manner would be his sexual companion during the remaining year.

N: That's one of the versions of its history, yes. But there are other ideas about where it came from. You are a girly girl and love all of that romantic stuff! You're not really interested in the history.

K: True, I am a romantic, but I think all girls are. It's a wonderful feeling to receive flowers from the one you love. Oh and chocolates too, and jewelry….

N: So, you just like getting gifts, then. I hope your boyfriend is rich! How many flowers do you expect? In England, it is traditional to give a dozen red roses.

K: A dozen? That's not enough, I expect 99 red roses. Now that is romantic, a dozen is not nearly enough.

N: OK, so what would you say if your boyfriend bought you a bunch of carnations or tulips or some other flower?

K: No, I wouldn't accept it. It must be roses. I guess Valentine's Day has become a really commercial festival here in China, maybe even more so than in the West, especially for young people.

N: How about Chinese Valentine's Day? I read that you have your own Valentine's Day, so why do you need to celebrate the Western one too?

K: As I said, Western Valentine's Day is commercial, and a good excuse to have your boyfriend buy you anything you want and take you out for an expensive dinner. Whereas Chinese Valentine's Day or "Qi Qiao Jie" is more traditional, it's based on a beautiful story.

N: Yes, I read something about that. Chinese Valentine's Day is about a love between the 7th daughter of the Emperor of Heaven and an orphaned cowherd.

K: Yes, that's right. The Emperor separated them. The 7th daughter was forced to move to the star Vega and the cowherd moved to the star Altair. They are allowed to meet only once a year on the 7th day of the 7th lunar month.

N: You certainly know your Chinese traditions!

N: 你最喜欢哪个西方的节假日呢？我敢打赌是圣诞节吧，圣诞节在中国非常受 30 岁以下的年轻人欢迎。

K: 我认为是情人节。它有一段非常有意思的历史，在 800 年以前，当时还没有情人节，罗马人在二月的月中设立了异教徒庆典。庆典的一部分内容就是抽签，年轻男子从一个盒子中抽出某个少女的名字，那么这个女孩随后就将成为该男子在这一年中的性伴侣。

N: 这是有关情人节来历的一种说法。但还有其他有关情人节来历的说法。你是个少女，热爱所有浪漫的食物，你不会对它的历史感兴趣的。

K: 没错，我是个浪漫的人，但我认为所有女孩都是浪漫的。收到你所爱的人送的鲜花，这种感觉实在是很棒。对了，还有他们送的巧克力、珠宝首饰……

N: 你仅仅是喜欢得到礼物，但愿你的男朋友很富有。你希望能得到多少花呢？在英国传统上是送 12 支红玫瑰。

K: 12 支？不够，我希望 99 支红玫瑰。那才够浪漫呢，12 支太少了。

N: 如果你的男朋友给你买了一束康乃馨或者郁金香或别的花，你会怎么样？

K: 不，我不会要的。必须是玫瑰花。我想情人节在中国已经成为一个真正的商业性的节日了，可能甚至超过了西方，尤其是对年轻人来说。

N: 那中国的情人节怎么样呢？我知道你们有自己的情人节，为什么你们会庆祝西方人的情人节呢？

K: 就像我所说的，西方的情人节是商业性质的，是一个很好的借口让你的男朋友给你买任何你想要的东西，以及带你去吃顿大餐。但是中国的情人节也就是"七夕节"，是比较传统的，它源于一个很美丽的故事。

N: 是的，我读过一些关于它的事情，中国的情人节是一个发生在玉皇大帝的第七个女儿和放牛的孤儿之间的爱情。

K: 是的，没错。玉皇大帝拆散了他们。七仙女被放逐到了织女星，牧童被放逐到了牛郎星。他们只被允许在每年的农历 7 月初七见一次面。

N: 中国的这些传统故事你当然知道了。

Questions 2

1. Which Western holiday or festival do you like the best? Why?
2. What do you expect from your boyfriend on Valentine's Day?

Dialogue 3

Kate: Nick, I hope you can help me to understand something. I was chatting with another foreign friend and I asked her about Christmas and she started talking about something called "Boxing Day". Are British people really so violent? I mean I've heard of football hooligans....

Nick: Kate, you've **got the wrong end of the stick**①. "Boxing Day" is the day after Christmas Day and has nothing to do with sports. British people don't go around beating each other up on the 26ᵗʰ December, you know.

K: So, why is it called "Boxing Day" if it's nothing about boxing? Why give it such a silly name. Or maybe it has something to do with The Boxer Rebellion?

N: No, no. Let me explain, Boxing Day began in England, in the middle of the nineteenth century, under Queen Victoria. Boxing Day, also known as St. Stephen's Day, was a way for the upper class to give gifts of cash, or other goods, to those of the lower classes. Maybe a member of the merchant class would give boxes of fruit or clothes to their servants.

K: I see. So, it's only celebrated in Britain? I've never heard an American talk about it.

N: You're right, Americans don't celebrate it. Boxing Day is celebrated in Australia, Britain, New Zealand, and Canada.

K: But it does have something to do with the word "box" not "boxing", because the gifts the masters gave to their servants were in boxes?

N: That is one theory, yes. Another is that Boxing Day comes from the tradition of opening the alms boxes placed in churches over the Christmas season. Which again were distributed amongst the poor, this time by the clergy, the day after Christmas.

K: That seems very old fashioned. What do people do nowadays? I know you are not very religious and only go to church once a year on Christmas Eve, so I can't expect you to do this.

N: That's right. Today, Boxing Day is spent with family and friends with lots of food and sharing of friendship and love. Government buildings and small businesses are closed but the malls are open and filled with people exchanging gifts or **maxing out**[2] their credit cards in the sales.

K: So some sales start on Boxing Day? Not on January 1st? I've heard of January Sales but never Boxing Day sales.

N: Many sales start on Boxing Day and the shoppers can be over serious. They run around everywhere, trying to find the best bargains. It's a good way to start exercising off all of the extra weight we all have gained after eating so much!

习惯用语 3

① get the wrong end of the stick：完全误解了
② maxing out：极度花费，尽情消费

K: Nick，你能帮我看一下吗，我在跟另一个外国朋友聊天，我问她关于圣诞节的事情，她说了一些关于一个被叫做"Boxing Day"的事情。英国人都这么暴力吗？我曾经听说过有足球流氓……

N: Kate，你误解了她的意思。Boxing Day 是圣诞节后的第一天，与任何运动丝毫没有关系。英国人不会在 12 月 26 日出去到处对打的。

K: 那如果那天没有任何关于拳击的活动又为什么叫做"拳击日"呢？为什么起这么个傻名字。也许它是从"拳击对抗赛"而来？

N: 不是的。我给你解释一下。Boxing Day 起源于英国，在 19 世纪中叶，维多利亚女皇统治时期。Boxing Day 也叫做圣·史蒂芬日，是上流阶层将钱币或其他物品作为礼物给那些下流阶层的日子。很可能是商人将成箱的水果或衣服给他的扑人。

K： 我明白了。它只是在英国有吗？我从来没有听美国人说起过。

N: 是的，美国人没有这个庆祝。Boxing Day 只在澳大利亚、英国、新西兰和加拿大有。

K: 它是由单词"盒子"而来，并非"拳击"，是因为主人给他的仆人们的东西都是放在盒子里的？

N: 是的，那是其中一个原因。另外一个由来是在圣诞节期间，有这样一个传统，人们将救济品放在盒子里摆放在教堂，由牧师将它们分发给穷人，也是在圣诞节之后的天。

K: 那似乎是个非常老的习俗了。如今人们都做什么？我知道你不是虔诚的信徒，也只是在每年的圣诞节前夜去一次教堂，所以你是不会做这种事情的。

N: 是的。现如今，在 Boxing Day 人们会跟家人和朋友共度，共同分享食物、友谊和爱。政府机构和小商铺会关门，但是购物中心会营业，里面都是互换礼物的人们，或者用他们的信用卡在特卖场尽情消费。

K: 那么在 Boxing Day 会有一些折扣吗？不是在 1 月 1 日？我听说过"一月甩卖"，但是从没听说 Boxing Day 的打折。

N: 许多折扣活动在 Boxing Day 开始，购物者也是非常重视的。他们会跑遍各个角落，试图找出最实惠的东西。这可以说是减掉身上肥肉的一个很好的方法，因为在大餐之后我们都长了太多肉。

Questions 3

1. How many U.K. holidays and festivals can you name?
2. Do you have any special times of the year when there are shopping sales?

Leisure 休闲

Pets

35 宠物

Background Information

In China all dogs require a licence and a rabies vaccination. In Beijing the licence is 5,000 yuan for the first year and 2,000 yuan annually after that. A home can only have one dog and dogs are not allowed in some public places like supermarkets. In Beijing they are only allowed outdoors on a leash between 8 pm and 7 am.

在中国养宠物狗都要有一个许可证，都要接种狂犬疫苗。北京的许可证，第一年交 5000 元，从第二年起每年交 2000 元。一家只允许养一只狗，并且不允许带到超市这种公共场合。在北京只可以在晚 8 点到早 7 点期间在狗被栓着的情况下出门遛狗。

背景信息

Dialogue 1

Sunny: Did you have any pets when you were young?

Nick: I had a dog, some rabbits and some goldfish.

S: Which is Britain's most popular pet?

N: I would have to say dogs. In fact, we often say that a dog is a man's best friend.

S: We have dogs as pets here too.

N: Yes, but most of your dogs seem to be very small. Ours would be a lot bigger and we would have a wider variety of breeds.

S: We prefer smaller dogs because they look cute and they're also easier to keep in our apartments.

N: One thing that you do that we would never do is eat them.

S: We think, especially in winter, that eating dog is good for our health. Have you ever tried it?

N: I tried it once but I kept trying to think what breed of dog it was and what part of the dog it was and so my appetite was ruined.

S: So you've never tried it since?

N: No! And I've never told my mum either because she's a dog lover!

S: 你小时候养过什么宠物吗?

N: 我养过一条狗、几只兔子和几条金鱼。

S: 英国人最喜欢养什么宠物呢?

N: 我觉得应该是狗吧。实际上,我们经常说狗是人类最好的朋友。

S: 我们这儿也养狗当宠物。

N: 是,但是你们养的狗似乎都很小。我们都喜欢养大型犬,而且狗的品种丰富一些。

S: 我们更喜欢小型犬,因为看起来很可爱,也更方便在房间里养。

Leisure 休闲

N: 你们做的一件事是我们永远不会做的，就是吃狗肉。

S: 我们认为，特别是在冬天，吃狗肉对健康有好处。你从没吃过？

N: 我曾经吃过一次，但是我会一直在想，这是什么品种的狗，吃的是狗的哪个部位的肉，这么一想就特别倒胃口。

S: 所以从那以后你就不再吃了？

N: 不吃了！而且我从来没跟我妈妈说过，因为她特别喜欢狗！

Questions 1

1. What pets did you have when you were young?
2. Do you ever eat dog? Why? Why not?

Dialogue 2

Trisha: Gross! Look at this, there is doggy business in the garden again.

Vincent: What do you mean, business? Dogs do business?

T: I'm talking about when dogs go to the toilet. It seems that in these apartment buildings no one cleans up after taking their dogs out.

V: But why should they?

T: It's **a common courtesy**①. In England, if the police see you letting your dog go to the toilet in a public place, and you don't clean it up, you can get **an on the spot fine**②.

V: Well, in this community there are cleaners who clean the garden every morning. It's included in the management fee. Maybe we just came at the wrong time of the day.

T: Well, I think dogs are annoying. Everyone keeps those small, yappy ones, they yap and yap and keep me awake.

V: Some people keep big dogs. In fact, they are becoming more and more popular. Personally, I like dogs and think they make great companions. I'd really like a husky.

T: A husky? But your apartment is so small! How could you keep a huge husky in there? I think you are cruel. I may not like small dogs, but I can see the logic behind keeping them rather than big dogs.

V: So, in Western countries does everyone keep a big dog? Your houses are all so big. But I don't see why dogs need so much room, as long as you love them enough, it's all that matters.

T: Yes, if we have a big house, we will keep a big dog and take it on plenty of walks. English people don't like horrible, little, yappy dogs as much. And most of all, we will clean up after it when it uses the toilet. English people are known for loving their pets.

V: I have heard the saying that dogs are man's best friends, but I think you just feel that little dogs are a menace!

习惯用语2

① a common courtesy：公共礼仪
② an on the spot fine：现场交纳的固定罚金（当违反法律的时候必须马上支付的钱）

T: 哦！看，在花园里又有"小狗买卖"了。

V: 什么意思，买卖？小狗能做买卖？

T: 我的意思是说狗屎。这些公寓的人遛狗之后，有狗屎了却好像没人清理。

V: 他们为什么要清理啊？

T: 这是社会公德心的问题。在英国，如果警察看见你让你的狗在公共场合大小便，而你不清理，那就会罚款。

V: 嗯，这个社区每天早上都有保洁员清理花园，这已经包含在管理费里了。可能我们刚好来得不是时候。

T: 嗯，我觉得狗太吵了。每个人都养那种小型的爱闹的狗。他们一直叫，吵得我都睡不着觉。

V: 有些人养大狗。实际上，大狗越来越受欢迎。从个人而言，我喜欢狗，认为它们是好伙伴。我特别喜欢哈士奇。

T: 哈士奇？但是你的公寓太小了！你怎么养那么大一只哈士奇呢？我觉得你太残忍了！我可能不喜欢小型犬，但是我知道，房间小的话，养小型犬比养大型犬更合适。

V: 那么，在西方国家是不是人人都喜欢养大型犬？你们的房子都那么大。但是我不知道狗为什么需要这么大的空间，只要你爱它们，这才是最重要的。

T: 是的。如果我们有个大房子，我们都会养大型犬，带它散步，英国人不喜欢令人讨厌的、到处乱叫的小狗。最重要的是，小狗大小便后我们都会清理干净。英国人以爱宠物著称。

V: 我听过这样一句话，狗是人类最好的朋友，但是我觉得你却认为小狗很让人讨厌。

Questions 2

1. What kind of dogs do you like best?
2. Would you keep a big dog in a small flat? Why? Why not?

Dialogue 3

Nick: All I ever see are pekingnese! Don't Chinese people like other breeds?

Sunny: We do! It's just that we have a special affection for pekingnese because they are **a miniature version**[①] of the ancient "Fo Dogs" of China which, because of their terrifying lion-like appearance, were thought to ward off evil spirits.

N: I don't think they're very frightening, especially being so small. And anyway, most of their owners carry them about in their arms. I think the owners get more exercise than the dogs!

S: We look after them well because they used to be favourites of the Chinese Imperial court where they were bred in great numbers but only those within the royal circle were permitted to own one. The dogs were lavishly tended, considered to be bringers of good fortune and possessed of the courage of the lion.

N: So you say but I want to hear proof!

S: OK. After British troops stormed the Summer Palace in Peking in 1860 there they found five small dogs protecting the body of the Imperial Princess who had taken her own life. So that proves they are brave and courageous.

N: But they didn't bring any good fortune to the Imperial Princess, did they?

S: Well, perhaps not but they did bring good luck to your Queen Victoria.

N: How?

S: Well she reigned for another 41 years so that was good fortune for her.

N: I suppose so. Our royal family really like dogs, especially corgis which are small dogs. I suppose your pekingnese started a trend for what we call toy dogs.

习惯用语 3

① a miniature version：小号版本，微缩型号

N: 我看见的都是京巴！中国人不喜欢别的品种吗？

S: 我们喜欢！只是我们对京巴有特别的感情，因为它们是中国古代"福狗"的微缩版，由于他们有狮子般的恐怖外表，所以认为它们可以挡住邪气。

N: 我没觉得有多恐怖，特别是这么小就更不觉得恐怖了。不管怎样，大部分养狗的人都是把狗抱在怀里来遛狗，这样我觉得主人比狗锻炼得更多！

S: 我们很好地照顾他们，因为他们是中国皇家最喜爱、养得最多的狗，而且只有皇亲国戚才允许养这种狗。这种狗深受人们呵护，

认为它能带来好运，并且具有狮子的勇气。

N: 这都是你自说自话的，但我想看到证据！

S: 证据有啊。早在 1860 年，在英国军队席卷了北京颐和园之后，他们发现了 5 只小狗保护着一位皇家公主的尸身。那就证明了它们的勇敢和胆量。

N: 但是他们没有给那位公主带来好运，不是吗？

S: 嗯，可能没有，但是给你们的维多利亚女王带来了好运。

N: 怎么回事？

S: 自那以后她又统治英国长达 41 年，所以给她带来了好运。

N: 我想是的。英国王室确实很喜欢狗，特别是小型的威尔士矮脚狗。我想你们的京巴开创了一个玩具狗的潮流时代。

Questions 3

1. Do you prefer small dogs or big dogs? Why?
2. Do you think that some dogs can bring good fortune to their owners?

36 婚礼

Weddings

❙❙ **Background Information** ❙❙

The average amount spent by Chinese newlyweds living in cities on their wedding and residence was 120,000 yuan according to a 2007 China Wedding Expo report. 80 percent was spent on housing.

In the west it is traditional for the bride to wear white which symbolises virginity and purity. However, the bridegroom wears black which is a sign of mourning!

2007 中国婚礼研究报告指出，中国城市的新婚夫妇平均花费在婚礼和住房上的总费用是 12 万元，其中 80% 花在住房上。

在西方，按照传统新娘要穿白色的婚纱，因为白色象征着纯真。而新郎要穿黑色的礼服，以象征悲哀（快乐的单身生活结束啦！——译者注）。

背景信息

Dialogue 1

James: Hey, Helena. Here are the wedding photos you wanted to see, sorry I didn't bring them in sooner.

Helena: You both look so wonderful in your wedding photos. Your wife, in particular looks stunning.

J: Yes, she does looks pretty amazing, doesn't she? I guess I look acceptable because of the good lighting and all of the make-up they put on me!

H: You both look great. Anyway, I was wondering why you have so many different types of photos? I mean, this one here, is outside in a park and you are both wearing traditional, Western style outfits. But in this one you look like something out of a Peking Opera play!

J: But we choose many different ones so that we can show them to everyone at the wedding.

H: At the wedding! But we only take the photos after the wedding, not before!

J: Well, that's strange to us. Anyway, in this picture you can see my wife in a traditional Chinese Qi Pao.

H: I think it's beautiful. I'd love to wear one when I get married, especially a red one.

J: Actually, I think that will be possible as it seems that Eastern styles, especially Chinese ones are becoming more and more popular in the West. Why don't you get a Qi Pao made while you are here, then you can wear it when you and Tony get married.

H: That's an excellent idea, James. I think I will.

J: 嗨，Helena，这是你想看的婚礼照片，不好意思没有早点儿带给你。

H: 你们俩在结婚照里看起来都太棒了。你妻子实在是太迷人了。

J: 是的，她确实看上去非常迷人，对吧？我想我看上去也还不错，这都是因为灯光和化妆。

H: 你俩看起来都很棒。不过，我还是想知道为什么你们要照这么多种不同类型的照片？我的意思是这张是在公园里，你们都穿着传统西式风格套装拍的，而这张你们看上去像在演京剧！

J: 我们选择许多不同风格的照片是为了能在婚礼上展示给每一个人看。

H: 在婚礼上！但我们是在婚礼后才拿到照片，而不是之前！

J: 那对我们来说是很不可思议的。不管怎样，这张照片你能看到我妻子穿着传统的中国旗袍。

H: 我觉得它很美，我结婚的时候也想穿一件，特别想要件红色的。

J: 实际上，我想它会很有东方格调，特别是中国的东西在西方越来越流行。为什么你现在不做一件旗袍呢，然后当你和 Tony 结婚的时候你就可以穿了。

H: 这个主意太棒了，James。我想我会的。

Questions 1

1. Which clothes would you choose for your wedding, Western or Chinese style?
2. Why do you have your photos taken before the wedding, do you think?

Dialogue 2

Anna: I've been invited to a wedding next weekend and I have no idea what to wear or to give as a gift. Could you give me some advice?

Vicki: Of course, even though I'm not married yet, I've been to tons of weddings.

A: So, first things first. What am I going to wear? Should I be dressy, formal, casual … or what?

V: To be honest, not everyone dresses up too much, so I would go for smart-casual. Somewhere in the middle.

A: What about style? I mean, should I wear a Qi Pao?

V: I wouldn't recommend it! Another of my foreign friends came to a wedding with me last year and wore a Qi Pao. Apart from the bride and waitresses, she was the only person wearing one. And hers was very bright blue and eye catching!

A: Oh dear. I bet she **stuck out like a sore thumb**①! I will not be wearing a Qi Pao, then.

V: OK, on to gifts. It's traditional to give something made from jade, perhaps a jade Chinese leaf. It symbolises health and wealth.

A: You mean a vegetable! Haha! That seems like a strange gift to me. Is there anything else I could give? Like household appliances? In England, we usually give something like a toaster.

V: I wouldn't recommend that, they more than likely already have everything they need for their new house from the family. If I were you, I would give the most traditional of all gifts, a red envelope filled with money!

A: That sounds like a good idea. But how much should I give them? I don't want them to think I'm showing off my foreign wealth, but I don't want them to think I'm cheap, either.

V: Well, it depends on how well you know them. If they are close friends, I would say you should give around 1,000 yuan, but if you don't know them so well, a couple of hundred would be enough.

A: I've known them for a long time and we are really close! Looks like it's going to be an expensive day for me!

习惯用语 2

① stuck out like a sore thumb：显眼，易被认出（由于一些奇特的或特有的特征）

A: 下周末我被邀请去参加一个婚礼，我不知道穿什么去，也不知道送什么礼物合适。你能给我点儿建议吗？

V: 当然可以，虽然我没结过婚，但我参加过许多次婚礼。

A: 这样啊，那第一个问题，我应该穿什么？我应该穿得讲究点儿、礼服、便装……还是其他的什么？

V: 坦白地讲，不是所有人都很讲究，所以我会穿时尚些的便装，不要太正式，也不要太随便了就行。

A: 那款式呢？我的意思是我能穿旗袍吗？

V: 我不建议你穿旗袍！去年我的另一个外国朋友跟我参加一个婚礼就穿了一件旗袍。当时，除了新娘和女服务员外，她是唯一一个穿旗袍的。她穿的是一件亮蓝色旗袍，很引人注目。

A: 天啊，那一定是有点儿喧宾夺主了。那我就不穿旗袍了。

V: 好的，关于礼物。传统上是赠送一些用玉做的东西，例如中国的玉叶。它象征着健康和富有。

A: 你的意思是蔬菜啊！哈哈，这种礼物对我来说太奇怪了。有其他我可以送的吗？像家用电器？在英国，我们通常送一些家用的东西，例如烤面包机。

V: 我不建议送那种东西，他们多数都会准备好新家需要的所有东西。如果我是你，我会准备最传统的礼物，用红色信封装一些钱。

A: 这个主意倒不错。但是我应该给多少呢？我不想让他们觉得我炫耀我的富有，但也不希望让他们觉得我穷。

V: 这个取决于你与他们的熟悉程度。如果他们是你的密友，我觉得你应该给 1000 元，但是如果你跟他们不是很熟，200 元就应该足够了。

A: 我认识他们很久了，而且我们关系也非常好。看来那天我要花很多钱了。

Questions 2

1. What kind of gift would you give at a wedding?
2. What kind of clothes would you wear?

Dialogue 3

Anna: So, what's going to be happening today? We've all arrived at the hotel and had our photos taken, what's going on next? It's totally different to England. Usually we have a ceremony in a church, for friends and family. Then later in the day we go somewhere else for a big meal and then a disco.

Vicki: Yes, today's going to be a new experience for you. Well, as Chinese are not usually religious, some people don't have any ceremony. But today, there will be a ring giving ceremony in the banquet room. It's popular to do this nowadays.

A: And do they say vows or "I do" or anything like that? Obviously they won't swear on God's name, but what do they say?

V: They do have some vows, yes, where they say how much they love each other. Then they do say "I do".

A: What happens after the ring giving ceremony? You said it takes place in the banquet room, so does everything happen in the same place?

V: First the ceremony, then the banquet and lots of toasts. There are always many bottles of expensive Chinese alcohol on each table. The groom must toast with everyone, so sometimes, to avoid getting totally drunk he will fill his glass with water, but everyone pretends they don't know! I heard that in England you do the toasts after the meal, right?

A: Some people do it after, some before. But we don't do it during the meal. So, what can I expect to eat today? Lots of yummy things I guess!

V: You must have a chicken dish, as chicken represents the phoenix. Also, there must be some fish because fish means plentiful. You should also get to sample some expensive seafood because the bride and groom give this to show that they appreciate how much money you put in your red envelope to them.

A: I see. Anything else? Someone told me something about lion's heads that I don't really understand. I mean, I can't even eat a duck head let alone a lion head!

V: Not a real lion's head! Lion's head meatballs. We use the name "lion's head" because they are so big. There must be four of these per table.

A: Why four? Why not two or three? I thought four could mean death! We certainly don't want to wish death on the bride and groom.

V: Not in this situation. Here, four means "double, double happiness". You know the slogan for all weddings is "double happiness".

A: 今天接下来要干什么？我们已经到饭店了，也拍完照片了，之后呢？这与英国完全不一样了。通常我们跟家人和朋友在教堂举行仪式。之后我们会去找个地方吃顿大餐，然后蹦迪。

V: 是的，今天你将有一个新的经历。因为中国人通常不信教，所以有些人不会举行任何仪式。不过今天，在宴会厅会有一个新郎新娘互赠戒指的仪式，这是现在很流行的。

A: 他们会发誓或说"我愿意"或其他类似的话吗？显然他们不会以上帝的名义发誓，那他们会说什么？

V: 是的，他们也发誓，说他们相爱彼此有多深，之后他们说"我愿意"。

A: 这个互赠戒指的仪式之后是什么？你刚才说是在宴会厅举行，那所有活动都在同一个地方举行了？

V: 首先是仪式，然后是宴会和祝酒。每桌上都有许多瓶非常昂贵的中国酒。新郎必须给每个人敬酒，所以有时候新郎为了避免喝醉酒，他会把杯子里倒上水，但每个人都假装不知道。我听说在英国你们是在饭后敬酒，是吗？

A: 有的人在饭后敬酒，有的人在饭前敬。但是我们决不会在进餐的过程中敬酒。今天都会有什么吃的呢？我想一定有许多美味吧！

V: 一定得有盘儿鸡，因为鸡代表的是凤凰。也一定有鱼，鱼象征富裕。还会有些很贵的海鲜，这是新娘和新郎对你们送给他们红包表示感谢。

A: 我知道了，其他的呢？有人跟我说过狮子头，我没明白是怎么回事。我的意思是，我连鸭头都不敢吃，更别说狮子头了。

V: 不是真的狮子的头，是肉丸子。我们之所以叫"狮子头"是因为它太大了。每桌必须是四个。

A: 为什么是四个？为什么不是两个或者三个？我觉得"四"意味着死！我们肯定不希望新郎新娘死吧。

V: 不是这样的。这里，"四"意味着"双喜"的意思。你知道所有婚礼的标语都是"双喜"。

Questions 3

1. What kind of ceremonies have you seen at weddings?
2. What kinds of food would you expect to eat at a wedding?

37 佛教

Buddhism

Background Information

Buddhism entered China a few centuries after the passing away of the Buddha, at a time when Confucianism and Taoism were the predominant religions in a country that was as big as a continent and rivaled India in historical antiquity and cultural pluralism. In the early phases of its entry, Buddhism did not find many adherents in China. But by the 2nd Century A.D., aided to some extent by the simplicity of its approach and some similarities with Taoism, it managed to gain a firm foothold and acquired a sizeable following.

The arrival of many new Buddhist scholars from the Indian subcontinent and central Asia, gave an impetus to the new religion that had many attractive features besides an inbuilt organisational approach to the study and pursuit of religion. During the same period many Buddhist texts were translated from Pali and Sanskrit into Chinese.

佛教在佛祖仙了几个世纪之后进入中国。当时，在这样一个幅员辽阔的国家，像印度一样具有悠久的历史和多元的文化，儒教和道教是占据主导地位的宗教。在佛教进入中国的早期，没有多少信徒。但是到了公元2世纪的时候，通过简化途径和模仿一些道教的内容，佛教增加了一些范畴，从而设法取得了一个坚实的立足点，同时拥有了相当多的追随者。

从印度次大陆和中亚来的许多新佛教的学者，给了这个新的宗教一个推动力，除了一个内在的有组织的研究和对宗教信仰的追求，它还具有许多吸引人的特点。在同一时期内，许多佛教典籍被从巴利文和梵文翻译成中文。

背景信息

Dialogue 1

Tori: So, is it true that Buddhism is a religion for some, and is a philosophy or a culture for others? That it is almost impossible to define, because there are so many kinds of Buddhism and so many contradictions within the overall traditions?

Crystal: Yes, that's right. But there are important **common threads**[①] such as the so-called "Three Jewels", you can follow.

T: What are they? Is this jewel thing a metaphor? Buddhism characteristically describes reality in terms of process and relation rather than entity or substance.

C: Not quite. They're all in Indian words, but mean the presence of the certain concepts of Buddha, proper behaviour and truth, and any group of devoted and spiritually committed Buddhists.

T: Oh, I know these three essential elements for Buddhists. First, of primary importance to concepts are the sense of selflessness achieved by way of inner searching, often in a monastic setting, and a goal of enlightenment, or nirvana. And I don't mean the American Grunge band. In Buddhism, nirvana is the ultimate place where all Buddhists are trying to find.

C: OK. So what else can you tell me, do you know the other two?

T: I think so. Second, in the Buddhist world, behaviour and truth can be found in the life and teachings of the Buddha. And the third is the holy community of Buddhists, a place of spiritual refuge. ·

C: Do you know there are the "Four Noble Truths" of the Buddha's teachings at the basis of the community?

T: I'm not sure, but I know the Eightfold Path, which contains the elements necessary to enlightenment.

C: Wow, great. You really do know a lot. Can I carry on? In the "Four Noble Truths" existence is suffering; suffering has a cause, namely craving and attachment; and when suffering stops, there is nirvana. To reach nirvana, you must suffer.

T: All right. It seems the "Eightfold Path" is very important. This seems to be the norm for all Buddhists, as I understand it.

C: Yes, these are the usual understandings. Just like in Christianity, when they treat others as you would wish to be treated yourself. With these ideals for living, you can't go wrong.

习惯用语 1

① common threads：相同的事情

T: 那么，佛教对一些人来说是一种宗教信仰，同时对另一些人来说是一种哲学或文化，是真的吗？这简直无法分辨，因为佛教有很多种，而且在全部的传统中还存在着许多矛盾。

C: 是的，正是如此。但你还是可以去追随重要的相同点，比如所谓的"三宝"。

T: 他们是什么？所谓的"宝"是一种比喻吧？佛教特征就是把现实的事物描述为过程和关系，而不是实体或实质。

C: 也不全是这样。"三宝"都是用印度语表示的，他们代表着佛、佛法（适当的行为和真理）和佛僧（任何在肉体上和精神上都忠于佛教的佛教徒组织）三个特定的概念。

T: 哦，我知道那三个对佛教徒至关重要的因素。第一，对这些概念最最重要的是通过自省来达到无我的境界，通常是在寺庙中修行，同时还要有一个教化的目标，就是涅槃。我不是指那个美国的摇滚乐队。在佛教的世界里，涅槃是一个终极的境界，所有的佛教徒都在致力于到达到这个境界。

C: 好，那你还能告诉我些什么呢，你知道其他两个"宝"吗？

T: 我想我知道。第二个是，在佛教的世界里，行为与真理都能在佛祖的生活中和传教中找到。第三个是佛教徒神圣的团体，它是一个精神庇护所。

C: 你知道在佛教界的基础理论中有个佛祖教导的"四圣谛"吗？

T: 我不太清楚，但是我知道那个"八正道"，它包含了达到涅槃的必要因素。

C: 哇，太好了。你知道的真多。我能继续我的话题吗？在"四圣谛"中，生既是苦；苦的因就是欲和痴；苦终则是涅槃。要到达涅槃的境界，你就得受苦。

T: 好的。"八正道"似乎也是非常重要的。按我的理解，它是所有佛教徒的行为规范。

C: 是的，这些就是通常的理解。就像在基督教中，对待别人就像希望别人如何对待自己一样。在生活中有这些美德，错不了。

Questions 1

1. What do you know about Buddhism?
2. Do you think that nirvana really exists? Why? Why not?

Dialogue 2

Tori: When Buddhism came to China, is it true that it was all in Sanskrit, the ancient Indian language? How did they change it, so locals could understand it? Didn't China already have its own religions?

Crystal: Yeah. In addition to this problem there was opposition from both the Confucian and Taoist schools of religious and philosophical thought. After being affected by Chinese culture, politics, literature, and philosophy, Chinese Buddhism originated one of the three major schools of thought along with Confucianism and Taoism. It became widely accepted.

T: But I read that at the beginning, most of the Chinese gentry were indifferent to the Central Asian travellers and their religion. Not only was their religion unknown, but much of it seemed alien and amoral to Chinese sensibilities. Concepts such as monasticism and individual spiritual enlightenment directly contradicted the core Confucian principles of family and emperor.

C: Yeah, Confucianism promoted social stability, order, strong families, and practical living. When people first started to hear about Buddhism, many Chinese scholars regarded it as merely a foreign branch of Taoism.

T: But it was not until the late fifth and early sixth centuries that Buddhism of a Mahayana sort was able to become a part of Chinese life, wasn't it? So by then it had become a spiritual complement to secular Confucianism and had provided the idea of Enlightenment to Taoism?

C: That's right. By that time, the three schools of thought would be seen as a complementary unity. Unlike in the beginning when everyone thought Buddhism was totally different.

T: Today the most popular form of Buddhism in both mainland China and Taiwan is a mix of the Pure Land and Chan School.

C: I know. Chinese Buddhism belongs to the "Larger Vehicle", right? I mean the emphasis should fall on the larger society rather than the individual salvation.

T: Yeah. The "Larger Vehicle" groups, represented by Mahayana, stress the ideal of the Bodhisattva as the person concerned with achieving Buddhahood. Ordinary people would find getting to this level an impossible struggle.

C: In contrast to that, the older Hinayana approach, termed by Mahayana as "Lesser Vehicle", stresses the ideal of the Arhat, the enlightened one who has attained nirvana.

Religion 宗教

265

Anyway, within the larger divisions of Hinayana and Mahayana, many diverse understandings and doctrinal divisions exist.

T: Ah, I know. The third school, the "Diamond Vehicle", the Vajrayana, has a long tradition in Tibet and Japan. I guess this school isn't so popular in China.

T: 当佛教传到中国时，所有的东西都使用古印度的梵文，是吗？人们是如何改变它，以便当地人可以理解它呢？当时中国还没有自己的宗教吗？

C: 有的。除了中国当时已有自己的宗教之外，佛教所弘扬的也和儒家和道家各派的宗教和哲学思想相对立。在经过中国的文化、政治、文学和哲学的影响之后，中国佛教开始成为与儒教和道教并列的三个主要的思想体系之一。佛教被广泛地接受了。

T: 但是我看过书上说，在刚开始的时候，大多数的中国贵族对于中亚的旅行者和他们的宗教无动于衷。不仅是因为他们的宗教前所未闻，而且它在许多方面给中国人的感觉似乎是异类的和反道德的。比如去庙宇修行和个人精神的教化等概念直接同儒家关于家庭和君主的核心原则相抵触。

C: 是的，儒教提倡社会稳定、守秩序、牢固的家庭观念，以及实际的生活。当人们刚开始听说佛教的时候，许多中国的学者只是把它当作道教的外国分支而已。

T: 但是直到5世纪末至6世纪初的时候，大乘佛教这一教派才能够成为中国人生活的一部分，对吗？因此直到那时，佛教对于提倡在家修行的儒教来说是一个精神上的补充，并且给道教提供了"教化"这个理念，是这样吗？

C: 对。当时，这三个思想体系被认为是互为补充的整体。不像刚开始的时候，每个人都认为佛教是完全不同的东西。

T: 今天，在中国大陆和台湾省，最普遍的佛教形式是净土宗和禅宗的结合体。

C: 我知道。中国佛教属于"大乘"佛教，对吗？我是说"大乘论"的重点是在于大的群体，而不是个人的救赎。

T: 是的。以"大乘佛教"为代表的"大乘论"教派强调菩萨是与成佛有关的个体。普通人会发现要达到佛的境界简直是不可能的。

C: 相反地，被"大乘佛教"定义为"小乘论"教派的，更早期形成的"小乘佛教"的佛法，强调阿罗汉是一个被教化了的个体，他达到了涅槃。总之，在更大的范畴里区分出来的"大乘佛教"和"小乘佛教"中，还存在着各种各样的理解和教派的分支。

T: 啊，我知道。第三个教派叫"金刚乘"，就是"密宗"，在西藏和日本有着由来已久的传统。我想这一教派在中国不是很流行。

Questions 2

1. Is there a Buddhist temple near you?
2. Have you ever been to it? Why?

Dialogue 3

Tori: From the beginning, meditation and observance of special morals were the foundation of Buddhist practice. The five basic morals, followed by members of monastic orders and **the laity**[①], are to refrain from taking life, so you must be a vegetarian. No stealing, acting unchastely, telling lies, and drinking intoxicants, you know things like alcohol and coffee.

Crystal: Yeah. Fasting is not central to Buddhist practice except for the monastic community, but dietary abstinence relates to a very widespread idea that giving up something desirable increases your spiritual side. If you give up something you love you suffer, it's a sort of test.

T: That's right. Monks are expected to show moderation and control in all things, including eating. They are warned that wrong mental states easily come to the surface when collecting or eating food.

C: Chinese Buddhism regulates communal meals as part of

monastic discipline. Monks used to grow their own food to provide vegetables for simple meals with rice or congee.

T: How do Chinese monks prepare their food nowadays? Especially, when for rituals or important events, what kind of food do they eat?

C: Rather than collecting food from begging or donations as in Theravada communities, Chinese monasteries often prepared food at the temple. Occasionally others might provide a vegetarian feast to celebrate Buddha's birth, enlightenment, and death, in order to gain merit.

T: OK. I think that Buddhism regards food as a metaphor, right? I mean eating is important for life.

C: That's right. Food as an object of meditation is a metaphor for the foulness of the body. The feeling of craving something is bad, it can make us weak if we give in. When Buddhists can gain this understanding, they can use this in their daily lives and when they meditate.

T: But not all Buddhists are vegetarian, but we do believe vegetarianism and Buddhism **go hand in hand**[2].

C: In China, Zen cooking is a style of vegetarian cooking developed by Zen monks, and prepared as a spiritual exercise with attention to balance, harmony, and delicacy.

T: As the Buddhism population grows in the West, Buddhist vegetarian restaurants have **popped up**[3] in lots of places, offering devotees and strict vegetarians an opportunity to try real Buddhist food, which has been made without animals suffering or being killed.

C: That's great. In fact, eating can be a kind of meditation. You know the phrase, "waste not, want not", well we can use this here. Buddhists believe they should eat just enough of the right kind of things.

① the laity：俗人（以别于教士或僧侣）
② go hand in hand：密切关联地
③ pop up：突然地出现

T: 从一开始，冥想和遵守特殊的道德要求就是佛教徒修行的基础。在寺庙修行的僧人们以及俗家弟子们遵守着五个基本的道德，它们是，避免杀生，因此你必须吃素、不偷盗、不淫欲、不说谎、不喝能上瘾的东西，比如说酒和咖啡。

C: 是啊。禁食除了适用于寺庙修行的群体之外，并不是佛教徒修行的中心，但是对饮食的节制关系到非常广泛传播的思想，即放弃一些渴望的东西以增加你的精神境界。如果你放弃你喜欢的东西，你就是受苦，这是一种测试。

T: 对呀。僧侣们应该在所有事情上都显示出适度和控制，包括吃饭。他们被告诫道，当乞食或吃饭时，错误的心智很容易地表露出来。

C: 中国佛教规定了共同进食作为寺庙修行的纪律。僧侣们通常种植自己的食物，以供给蔬菜、米饭或粥等简单的伙食。

T: 现在，中国的僧侣们如何准备他们的伙食呀？特别是在祭祀典礼或重要的活动时，他们吃什么饭呢？

C: 中国的寺院通常在庙里准备食物，而不像"小乘佛教"教派那样，需要通过乞讨或捐赠来采集食物。偶尔，普通人为了积累功德，也会准备一个素宴来庆祝佛祖的生日、开光，以及忌日。

T: 好的。我想佛教把食物看作是一种隐喻，对吧？我的意思是吃饭对生命来说很重要啊。

C: 对呀。食物作为冥想的对象是一种身体里的邪恶的隐喻。这种对贪欲的感觉是有害的，屈服于它会使我们脆弱。当佛教徒能够理解这些的时候，他们就能在日常生活中和冥想时加以运用。

T: 但不是所有的佛教徒都是素食者，可我们确信素食主义和佛教是密切相联的。

C: 在中国，禅膳是一种素食的烹调方式，是由禅宗的僧人发起的，并把它作为一种精神的锻炼，兼顾平衡、和谐和美味。

T: 在西方，由于佛教信众人数的增长，佛门的素食餐厅已经突然出现在许多地方，为皈依者和素食者提供一个尝试真正佛门膳食的机会，这样的食物不会使动物受害，以及避免了杀生。

Religion 宗教

C: 太好了。实际上，吃饭也能成为一种冥想。你知道有这样一句话，"俭则不匮"，我们可以用在这里。佛教徒相信他们应该只吃适量的正确的东西。

Questions 3

1. Are you a vegetarian? Why?
2. Do you know anything about Zen cooking?

38 道教
Taoism

Background Information

Taoism is a philosophy or way of life that may have been started by a man named Lao Tsu (or Lao Tzu) who lived a little before Confucius, about 600 B.C. Tao means the "way" or the "path". According to the traditional story, Lao Tsu worked as a librarian in the emperor's library (this was in the Eastern Chou dynasty).

Lao Tsu believed that the way to happiness was for people to learn to "go with the flow". Instead of trying to get things done the hard way, people should take the time to figure out the natural, or easy way to do things, and then everything would get done more simply. This idea is called "wu-wei", which means "doing by not doing".

道教是一种哲学或者生活方式，他是由一个叫老子的人发起的，老子生于公元前约 600 年，略早于孔子。道是"方式"或者"路"的意思。按传统的说法，老子是皇家图书馆的管理员（这是东周时期）。

老子相信让人们学会"顺其自然"就找到了幸福之路。与其强行去做某件事情，人们不如花些时间找出自然的或者从容的方法来做事情，那么任何事都会很简单地就办到了。这一思想叫做"无为"，意思是"不做而做"。

背景信息

Dialogue 1

Tori: I saw a strange guy riding a bike earlier. He looks like he came just off from a film set, wearing traditional clothes in grey colour, and put all his hair up into a pony tail on top of his head.

Gareth: Oh, he is probably a Taoist. You know the headquarters of Taoism is in Beijing. It's a famous Taoist monastery, which holds special markets every Chinese New Year.

T: Yeah, I've heard of that famous monastery before, but I've never been there.

G: Me neither. Anyway, it's a popular place for people who believe in Taoism or who just want to get some good luck.

T: Really? But I'm a little bit confused about Taoism. I heard Taoism is one of the three major religious systems of ancient China, together with Confucianism and Buddhism, but why are there fewer Taoist monasteries than Confucianism and Buddhism temples in China?

G: Taoism was officially proscribed in the 1950's in China, because throughout its history Taoism has provided the basis for many Chinese secret societies. Nowadays, Taoism is still practiced to some degree.

T: So that means in China, there are less Taoists than monks who live in monasteries. Also, what kind of lives do Taoists have? Are there still some secrets in their daily lives?

G: Not at all. They are ruled by a strict hierarchy, and a simple, ascetic life style is the norm in Taoist monasteries. The daily routine consists of several periods of seated meditation, worship, meals, and work in the gardens and the fields. Everybody is kept busy at all times, and all movements throughout the day are exactly prescribed and have to be done with utmost control.

T: They are just like other monastic religions. Is there anything else that makes them different from other religions?

G: Taoism has lots of liturgy and rituals, which makes the Tao more real to human beings and provides a way in which humanity can align itself more closely to the Tao to produce better lives for all. The religious elements of Taoism draw much of their content from other Chinese religions, including many local cults, and so enfold a very wide range of culture and belief within the wings of the Tao.

T: OK. Can I say that most Taoist temple practices are designed to regulate the relationship between humanity and the world of gods and spirits, and to organise that relationship, and the relationships in the spirit world, in harmony with the Tao?

G: That's right. You've got the main point, but Taoism is deeper than you or me can really understand.

T: 我刚才看见一个怪人骑着车。他看上去像刚从拍电影的片场跑出来，他穿着灰色的传统服装，并且把头发在头顶扎成了一个小辫儿。

G: 噢，他可能是个道士。你知道道教的总部在北京。它是一个著名的道观，每年春节都在那儿举办庙会。

T: 是的。我以前听说过那个著名的道观，但我从来没去过。

G: 我也没去过。总之，对那些信仰道教的人和那些只是想得到好运的人来说，它是一个受欢迎的地方。

T: 真的？但是对于道教我有点儿困惑。我听说道教是中国古代的三个主要宗教体系之一，与儒教和佛教相提并论，但是为什么在中国道教的道观要少于佛教和儒教的寺庙呢？

G: 在中国道教曾于 50 年代被官方禁止过，因为从道教本身的历史发展来看，道教曾经是许多秘密组织的基础。而今，道教还是在某种程度上被休习着。

T: 那么说在中国，道观里道士的人数要少于佛寺中的僧人啦。另外，道士的生活是什么样的？在他们的生活中还有哪些秘密呢？

G: 一点儿秘密也没有了。道士们被一个严格的等级制度管理着，而且一种简单的修行生活方式是所有道观的准则。日常生活包括几段时间的打坐、礼拜、用餐，以及在园中和地里耕作。每个人都是忙忙碌碌的，所有的行动都是被精确地指导的，并且要在绝对的控制下进行。

T: 他们就像其他在寺院里修行的宗教一样。道教还有什么区别于其他宗教的东西吗？

G: 道教有许多礼拜和典礼的仪式，这些仪式使得"道"对人类来说更加现实，而且这些仪式还提供了一种方式，这种方式使人性可以与"道"更紧密地结合在一起，从而为大家带来更好的生活。道教的宗教成分从其他中国的宗教中汲取了很多内容，包括许多当地的祭祀活动，并因此把一个广泛的文化和信仰的范畴包容在它的各教派之中。

T: 好的。我能这样说吗，就是大多数道观修行都是为规范人类与神灵世界的关系而设计的，并且用以组织人与神灵的关系，以及在神灵世界中的各种关系，使之与道和谐，可以吗？

G: 对了。你得到了要点，但是道教比你和我真正能够理解的还要深奥得多。

Questions 1

1. Have you ever been to a Taoist monastery? Which one?
2. Do you know anybody who is a Taoist?

Dialogue 2

Tori: I've heard that some parts of Chinese traditional medicine originally come from Taoism, is that true?

Gareth: Yes, it is. Also, Chinese alchemy, astrology, cuisine, Chinese martial arts, and many styles of breath training disciplines are intertwined with Taoism throughout history.

T: All right. Being one of the mainstream religions, but I can't really feel the existence of Taoism in China, because of the limited numbers of monasteries. But actually it affects our daily lives so much, even these days.

G: It's true. And not only does it give us certain life styles, it also shaped Chinese idealistic systems. We call it philosophical Taoism.

T: Philosophical Taoism? I thought Taoism is a kind of religion; it should be no more than the difference found in all religions between the practices of the faith, and the theological and philosophical ideas behind them. Is there a distinction between "religious" and "philosophical" Taoism?

G: Not exactly. Taoism was originally an esoteric philosophy, concerned with the unity underlying the opposites and diversity of the phenomenal world. Taoism taught union with the law of the universe through wisdom and detached action.

T: I see. It started with the thinking of the way the universe functions, or the path taken by natural events. Later, Taoism emphasized the techniques for realizing the effects flowing from that way or path, especially long life and physical immortality. Am I right?

G: Yes, you're right. The philosophical system stems largely from a text probably written in the middle of the third century B.C. It is characterized by spontaneous creativity and by regular alternations of phenomena that proceed without effort.

T: OK. That means philosophical speculation about what the way or path actually is should be very important.

G: Um. Not quite. The most important thing about the way or path is how it works in the world, and how human beings relate to it. The ideal state of being, fully attainable only by mystical contemplation, is simplicity and freedom from desire, comparable to that of an infant or an "uncarved block". You

Religion 宗教

275

know, something untouched and perfect.

T: I've got it this time. Human beings, following the way, must not feel greed or other normal human emotions. And everything must be moderated.

G: That's right. Taoist propriety and ethics emphasize love, moderation, and humility. Taoist thought focuses on non-action, spontaneity, humanism, relativism and emptiness.

T: 我听说中医里的一些东西最初是从道教中来的，是真的吗？

G: 是的。而且，中国的炼金术、占星术、烹饪、中国功夫以及多种呼吸练习法都在历史的发展过程中与道教纠缠在一起。

T: 好的。作为主流宗教之一的道教，因为道观数量少，在中国我不能真正感觉到它的存在。但是实际上，即使在今天，它也能如此广泛地影响到我们的日常生活。

G: 对呀。而且，不仅道教给予我们一定的生活方式，它还形成了中国的唯心论体系。我们叫它道家。

T: 道家？我认为道教是一种宗教；它的特殊性应该不会比见诸于各种宗教之间的修行活动的差别，以及它们各自理论和哲学思想的差别还要大吧。那么宗教的道教和哲学的道教有什么区别吗？

G: 也不完全是这样。道教最初是一种深奥的哲学，它涉及到一个感知世界的对立性和多样性的内在的统一体。道教通过智慧语言和超然的行为，用宇宙的法则来指导大众。

T: 我明白了。它以思考宇宙运行的方式，或者是自然活动的发展途径开始。后来，道教专著于一些技艺，这些技艺是为了认识到那些从以上方式或途径产生出来的影响的，特别是长寿和身体的不朽。对吗？

G: 是的，你说对了。这个哲学的体系大部分源于一本可能是写于公元前 3 世纪中叶的书。此书的特质就在于无意识的创造力和自然而然的规律性变化的现象。

T: 行啊。这就意味着关于用什么方式或途径的哲学思索应该是非常重要的。

G: 嗯，不完全如此。有关于方式或途径的最重要的事情是，它是如何在世界上发挥作用的，以及人类是如何受其影响的。只有通过神秘的沉思才能达到的理想的状态，就是质朴和从欲望中解脱，可以和婴儿或"璞玉"相比较。你知道，就是没有改变的和完美的东西。

T: 这次我终于明白了。按照这个方式，人类必须不能有贪婪或其他普通人类的情感。而且，任何东西都必须适度。

G: 对了。道家的礼仪和伦理强调爱、适度和谦逊。道家的思想专著于无为、自然发生、人道主义、相对主义，以及空无。

Questions 2

1. Have you ever read any books on Taoism?
2. Do you think that everything in moderation is good? Can you think of any exceptions?

Dialogue 3

Tori: I heard you've been practicing Taiji Kungfu for ages. Can you teach me some moves?

Gareth: Yeah, no problem. But you should learn some basic know-ledge about that, otherwise, the moves mean nothing. And you can never comprehend it.

T: I know some things about Taiji Kungfu. The full name is Taijiquan, and it's a part of the Taoists' practice. It embodies Taoist principles to a greater extent, and some practitioners consider their art to be a means of practicing Taoism.

G: OK. You're right. Generally, Taijiquan, with its fusion of ener-getic and relaxed exercise, has provided a means of increasing and enhancing Qi, the vital force of life. It's more like physical exercise than practicing Taoism nowadays.

T: Why is it called Taiji? Since it's about the body's energy and breath, Taijiquan and Qigong are related "internal" arts, so what does Taiji mean?

G: The Taiji is associated with Taoist symbolism, and represents the forces of Yin and Yang. It describes two primal opposing but complementary principles said to be found in all non-static objects and processes in the universe.

T: I see, but I didn't realize that deep meaning originated in ancient Chinese philosophy and metaphysics.

G: Well, the entire movement philosophy of Taijiquan is reflected by a long tradition of Taoists practicing exercises. Some of these were referred to Taoist Breathing. Exactly what these were and what their origins were is obscure but they are mentioned in Chinese chronicles as early as 122 B.C.

T: Wow, it's a long history. I thought Shaolin Kungfu was the oldest before.

G: Yes, Shaolin Kungfu was introduced in the sixth century A.D. , but Taijiquan was invented in the fifteenth century A.D. by a purported Taoist.

T: I think I've got it. Anyway, for its health, relaxation, and self-defence benefits, millions of people around the world practice it every day.

G: That's right. In any event the principles of yielding, softness, centeredness, slowness, balance, suppleness and rootedness are all elements of Taoist philosophy that Tai Chi has drawn upon in its understanding of movement, both in relation to health and also in its martial applications.

T: 我听说你曾经练过很长时间的太极功夫。你能教我一些动作吗?

G: 好的,没问题。但是你应该学一些有关的基本知识,否则的话, 动作说明不了什么。而且你永远也掌握不了它。

T: 我知道一些关于太极功夫的事情。他的全称是太极拳,它是道教 修行的一部分。它在更大的程度上体现了道教的原则,而且一些习拳的人认为他们的武术是一种道教修行的方法。

G: 是的。说得对。总之，由于太极拳融合了活力和放松运动，因此它提供了一种提高和加强生命的活力——气的方法。现在，太极拳更像是锻炼身体而不像是道教修炼了。

T: 它为什么叫太极？既然太极是有关身体的能量和呼吸的锻炼，那么它和气功是有关联的"内在"的技艺，因此太极到底意味着什么呢？

G: 太极和道教象征有关，而且代表着阴和阳两种力量。太极描绘的是两个最原始的相对且互补的要素，据说它们存在于宇宙中所有非静止的物体里和所有的过程之中。

T: 我明白了。我以前没有意识到这个起源于古代中国的哲学和玄学的深层次含义。

G: 是呀，整个太极拳动作的哲学反映了一个悠久的道教的修炼方法。有些动作涉及到道教的呼吸法则。到底是哪些动作以及它们的起源是什么已经说不清楚了，但它们在史书中最早的记载是于公元前 122 年。

T: 哇，有这么悠久的历史。我以前以为少林功夫是最古老的功夫。

G: 是的，少林功夫引进于公元 6 世纪，而太极拳是在公元 15 世纪时由一位传说中的道士发明的。

T: 我想我明白了。总之，由于太极拳的保健、放松和自卫的作用，每天世界上有成千上万的人在练习它。

G: 对呀。无论如何，太极的这些原则，像柔顺、温和、自我中心、缓慢、平衡、柔软、根深蒂固等，都是道教哲学的元素，太极已经在它的动作中描述出了对此的理解，这些动作都与健康和武术方面的应用有关。

Questions 3

1. Do you ever practice Taiji? Why?
2. Which is more important, the body or the mind? Why?

Religion 宗教

39   伊斯兰教

Islam

Background Information

    The most important Muslim practices are the Five Pillars of Islam which are the five obligations that every Muslim must satisfy in order to live a good and responsible life according to Islam.

The Five Pillars consist of:

- *Shahadah*: sincerely reciting the Muslim profession of faith
- *Salat*: performing ritual prayers in the proper way five times each day
- *Zakat*: paying an alms（or charity）tax to benefit the poor and the needy
- *Sawm*: fasting during the month of Ramadan
- *Hajj*: pilgrimage to Mecca

    伊斯兰教的五大支柱是最重要的穆斯林习俗，根据伊斯兰教的教义，它们是每个为了拥有美好的和可靠的生活的穆斯林必须满足的五项义务。

这五大支柱包括：

- Shahadah: 真诚地背诵穆斯林信仰的表白
- Salat: 每天五次以正确的方式进行祷告仪式
- Zakat: 施舍贫苦人
- Sawm: 斋月期间禁食
- Hajj: 麦加朝圣

背景信息

Dialogue 1

Nick: I'm interested to know how Islam arrived in China. Can you tell me?

Willa: Sure. It began with Muslim traders who were using the Silk Road. They did not deliberately promote Islam. It just happened naturally.

N: That was trade. What about contacts between the emperor and the caliph?

W: That happened in 650 A. D. and is considered to be the birth of Islam in China. A maternal uncle of Mohammed arrived in Guangzhou and the Emperor Yong Hui was so impressed by Islam that he ordered a mosque to be built.

N: Where was it and is it still there?

W: It's in Guangzhou and you can still see it today, even after 1300 years. It's called the Great Mosque of Guangzhou.

N: OK. So there must have been a lot of Muslims there. Where else did they settle?

W: Everywhere really. But most Muslims today will be in Xinjiang.

N: Why is that?

W: Xinjiang shares its borders with eight different nations, many of them with large Muslim communities so it was natural for them to migrate to Xinjiang.

N: I know that in Beijing there is a large mosque which is over a thousand years old. And of course you will usually find Muslim restaurants on campus as well as off.

W: Muslims today can be found everywhere. They are an integral part of China.

N: 我很想知道伊斯兰教是如何来到中国的。你能告诉我吗?

W: 好的。它始于利用丝绸之路的穆斯林商人。他们没有刻意宣扬伊斯兰教。它就自然而然地发生了。

N: 那是贸易往来。能谈谈中国皇帝和哈里发之间有什么交往吗？

W: 那是发生在公元 650 年的事，它被认为是伊斯兰教在中国的落户。穆罕默德的一个舅舅抵达了广州，而且当时的唐朝永徽皇帝对伊斯兰教的印象非常深刻，以至于下令建了一座清真寺。

N: 它在哪里呀，是否依然存在？

W: 它在广州，即使在 1300 年之后的今天你仍然可以看到它。它就是广州大清真寺。

N: 好的。那么那里一定已经有相当多的穆斯林。他们还在其他什么地方定居吗？

W: 到处都有，真的。但现在大多数穆斯林住在新疆。

N: 为什么呢？

W: 新疆和其他八个不同的国家接壤，其中多数国家具有大规模的穆斯林社区，所以他们很自然地迁移到新疆定居。

N: 我知道北京有一个很大的清真寺，它有一千多年的历史。当然啦，你通常会在校园内外发现穆斯林餐厅。

W: 今天穆斯林可谓比比皆是，他们是中国不可分割的一部分。

Questions 1

1. Have you ever seen a mosque? Where?
2. Do you like to eat in Muslim restaurants? Why? Why not?

Dialogue 2

Nick: How are Muslims regarded today in China?

Willa: Since 1978 when religious freedom was allowed Islam seems to have undergone a bit of a revival. There are scores of thousands of mosques in China with nearly 30,000 in Xinjiang.

N: But are they allowed to read the *Koran*?

W: Of course. They can read it not only in Arabic but also Chinese, Uygur and other Turkic languages.

N: Are they allowed religious holidays?

W: Muslim workers are permitted holidays during major religious festivals. Furthermore, they have also been given almost unrestricted allowance to make the Hajj to Mecca.

N: I remember seeing that on "China Today" on CCTV9.

W: In addition, where Muslims are in the majority the breeding of pigs by non-Muslims is forbidden in deference to their beliefs.

N: What about marriage?

W: They are allowed to have their marriages consecrated by an Iman and they also can have their own guburs.

N: That certainly is a great example of religious freedom!

N: 目前中国是如何对待穆斯林的?

W: 自从 1978 年宗教自由被获准后，伊斯兰教似乎已经历了一段复苏。在新疆有近 3 万座清真寺，在全国其他地方也有数千座清真寺。

N: 但允许他们看《古兰经》吗?

W: 当然可以了。他们不仅可以看到阿拉伯文的，也能看中文的，还有维吾尔语和其他突厥语族的版本。

N: 允许他们过宗教节日吗?

W: 回族劳动者被获准在主要的宗教节日放假。此外，他们还获得了几乎不受限制的假期额度以便去朝圣。

N: 我记得在中央电视台第九套的"今日中国"节目中看到过。

W: 此外，在穆斯林人口占多数的地区，非穆斯林人群是禁止养猪的，以尊重穆斯林的信仰。

N: 他们的婚姻情况如何?

W: 他们可以由伊玛目（伊斯兰教教职称谓，阿拉伯语意为领袖、师表、表率、楷模等——译者注）主持婚礼，而且他们还可以有自己的拱北（伊斯兰教术语，指伊斯兰教先贤的陵墓，一般指被特别装饰的坟墓——译者注）。

N: 那肯定是一个宗教自由的伟大榜样!

1. Do you have any Muslim friends?
2. Do you think Muslims should be allowed greater religious freedom? Why? Why not?

Dialogue 3

Nick: We've talked about the start of Islam in China and its present day role but what about Islam in Chinese history?

Willa: I think it was Genghis Khan who first employed Muslims at his court. He was always on the lookout for talented people no matter their religion or country of origin.

N: I would have done well under Genghis Khan! So what areas were they good at?

W: They virtually dominated the import/export business by the time of the Song dynasty (960 – 1279 A.D.). Indeed, the office of Director General of Shipping was consistently held by a Muslim during this period. Under the Ming dynasty (1368– 1644 A.D.) generally considered to be the golden age of Islam in China, Muslims gradually became fully integrated into Han society.

N: But because they were powerful weren't they disliked by the Chinese? Just like the Jews were hated in the West.

W: In spite of the economic successes the Muslims enjoyed during these and earlier times, they were recognised as being fair, law-abiding, and self-disciplined. For this reason there was little friction between Muslim and non-Muslim Chinese.

N: Are there any famous Muslims in Chinese history?

W: Perhaps the most famous Muslim in Chinese history was Admiral Zheng He.

N: I've heard of him. A book was written by Gavin Menzies called *1421* and claimed that his ships discovered America before Columbus.

W: He made seven voyages to many different countries in huge ships which was a tremendous feat at that time.

N: But his name is Chinese and I have some Muslims as students and their names are Chinese, too.

W: Many Muslims who married Han women simply took on the name of the wife. Others took the Chinese surnames of Mo, Mai, and Mu—names adopted by Muslims who had the names Muhammad, Mustafa, and Masoud. Still others who could find no Chinese surname similar to their own adopted the Chinese character that most closely resembled their name—Ha for Hasan, Hu for Hussein, or Sai for Said, and so on.

N: 我们已经谈过伊斯兰教在中国的起源及其在当代的角色，那么谈谈中国历史上的伊斯兰教吧?

W: 我认为是成吉思汗首先在他的朝廷里雇用穆斯林的。他总是在寻觅人才，而不论其宗教或原籍。

N: 在成吉思汗的手下我也会干得很好! 那么当时他们擅长什么呢?

W: 事实上，早在宋朝 (公元 960 –1279 年) 的时候，他们已经主宰了进出口业务。确实，在此期间航运总管办公室就一直由穆斯林掌握着。到了明代 (公元 1368 –1644 年)，穆斯林渐渐地完全融入了汉人社会，这个时期被普遍认为是伊斯兰教在中国的黄金时代。

N: 但是，因为他们的权势，会不会使其他中国人不喜欢他们? 就像在西方，犹太人遭人记恨一样。

W: 尽管在当时和更早的时候，穆斯林就享有了经济上的成就，他们被认为是公平、守法和自律的。因此，在穆斯林和非穆斯林之间没有多少摩擦。

N: 在中国历史上有没有著名的穆斯林啊?

W: 商船队长郑和或许是中国历史上最有名的穆斯林。

N: 我听说过他。孟席斯写了一本叫《1421》的书，声称郑和的船队比哥伦布更早发现了美洲大陆。

W: 他乘坐大船进行了 7 次航行，到过许多国家，在当时这是一个壮举。

N: 但他的名字是中文的，我有一些学生是穆斯林，他们的名字也是中文的。

W: 许多穆斯林娶了汉族妇女后，就简单地使用他们妻子的名字。还有些人采用中文的姓氏，像莫、麦和穆等，而名字取自穆斯林人的姓名穆罕默德、穆斯塔法、马苏德。还有的人找不到与自己的姓氏相似的中文姓氏，他们就选择最近似的汉字组成他们的名字——哈是哈桑，胡是侯赛因，或者赛为赛义德，等等。

Questions 3

1. Do you know of any famous Muslims in Chinese history?
2. Do you have any Muslim friends or neighbours?

40 基督教

Christianity

‖ Background Information ‖

Christianity does not start with the birth of Christ over 2,000 years ago but goes back to the very beginning of time. The *Bible* begins with creation, not only of the world and of all life but also records the beginning of man, marriage, sex and nations. The *Old Testament* records the fall of man and the looking forward to the coming of the Saviour. The *New Testament* describes the birth, life, death, resurrection and ascension of Jesus Christ. It then tells the story of the early church and ends with a description of the end of the world when Jesus returns and judges everyone, living and dead.

基督教不是从两千多年前耶稣诞生时开始算起来的，而是追溯到最初的时间。《圣经》以创世纪开始，不仅是世界和所有生命的开始，而且记载了人、婚姻、性和民族的开始。《旧约全书》记录人类的堕落，并期待着即将到来的救世主。《新约全书》中描述了耶稣基督的诞生、生活、死亡、复活和升天。然后它讲述了一个关于早期教会的故事，并以世界末日的描述为结尾，到那时耶稣回到人间并对每个不论是活着的还是死了的人进行审判。

背景信息

Dialogue 1

May: I have heard a lot about the *Bible* but never read it.

Nick: Well, someone once said that it's the world's best-selling book but also the world's least read book!

M: Why is that?

N: Some people think that the *Bible* is difficult to read because they are used to the *King James Version* where some of the English is difficult but there are many modern translations available now.

M: You said that the *Bible* is the world's best-selling book. Why is that?

N: Several reasons. There are many great stories in it such as David and Goliath, Jonah and the Whale and the Good Samaritan.

M: Wasn't he the one who helped someone in distress when nobody else would?

N: That's right. Jesus told that story to emphasise that everyone is our neighbour and we have a responsibility towards them.

M: So Jesus was a great teacher as well as a good man.

N: He was actually more than that. He was fully man and God too.

M: I find that hard to believe. Isn't the *Bible* just a collection of myths and fables?

N: Some people think so but most Christians take it literally as God's word which is another reason for its popularity.

M: But God didn't write it.

N: About 40 different people wrote it over a period of 1500 years on 3 different continents and in 3 different languages but all the writers claimed to have been inspired by God.

M: I see. Can you tell me another reason?

N: Many people see it as the finest literature produced in English, even better than Shakespeare.

M: 我常听人说起《圣经》，但还从未看过。

N: 是呀，有人说过它是世界上最畅销的书，但也是世界上看的人最少的书！

M: 为什么这么说？

N: 有人认为《圣经》是很难理解的，因为他们看的是《詹姆士国王钦定本》，那时的英语很难懂。但现在有了很多现代文译本。

M: 你刚说到过《圣经》是世界上最畅销的书。为什么呢？

N: 有几个原因。在《圣经》里有很多伟大的故事，比如"大卫王和巨人歌利亚"、"约拿和鲸鱼"，以及"好撒马利亚人"。

M: 耶稣是在人们危难时唯一能提供帮助的人吗？

N: 没错。耶稣讲这些故事是想强调每个人都是我们的邻居，我们对他们都负有责任。

M: 所以耶稣是一位伟大的导师，也是一个好人。

N: 不仅如此。他还是个完善的人，是上帝。

M: 我觉得难以置信。《圣经》不就是一本神话和寓言集吗？

N: 有些人是这样认为的，但大部分基督徒把它完全当作上帝的语录，这也是它如此受欢迎的另外一个原因。

M: 但上帝并没有写《圣经》。

N: 在 1500 多年间，由大约 40 个人写成的《圣经》，他们来自 3 个不同的大陆，以 3 种不同语言书写，但所有的作者都声称是受到上帝的启发。

M: 我明白了。你还能告诉我另一个原因吗？

N: 很多人认为它是英国最优秀的文学作品，甚至比莎士比亚的作品还好。

Questions 1

1. Have you ever read the *Bible*?
2. Which stories do you like the best?

Religion 宗教

Dialogue 2

Nick: At Christmas many churches hold nativity plays about the birth of Jesus. Was he really born in a stable?

Tori: No one knows. Some think he was born in a cowshed and others in a cave. In either case there is nothing remaining in Bethleham now.

N: How long did he live there?

T: Just for a couple of years until his parents fled to Egypt to avoid persecution from King Herod.

N: I remember, Herod had been told that a king had been born here and so he determined to kill his potential rival although Jesus was only a baby.

T: Later Jesus came back to Israel after Herod died and the family lived in Nazareth.

N: There he had several brothers and sisters too.

T: That's right and then when he was thirty years old he began his public ministry around the Sea of Galilee.

N: That's why so many of his early disciples were fishermen like Peter and James and John.

T: And he did so many miracles like feeding the 5,000, turning water into wine and walking on water.

N: Not forgetting healing the blind, the lame and the sick. Although I think his greatest miracle was to bring Lazarus **back from the dead**[1].

T: He was a great teacher and many of the stories he told like the Prodigal Son and the Lost Sheep have been **enjoyed by young and old**. [2]

N: I think the greatest tragedy was that most of his own people, the Jews, never believed that he was the Son of God.

T: The Jews were expecting the Messiah to come but they

expected someone who would rid them of their Roman conquerors not one who would be crucified on a cross.

N: We Christians believe that Jesus died for all our sins and that one day he will return to put an end to all wars and suffering.

T: The Jews believe that Jesus was a great prophet but that to call himself God was blasphemy.

N: I understand that. I guess the best thing to do is to understand each other's religion and agree to disagree agreeably, right?

T: Well put!

习惯用语 2

① back from the dead：使先前死去的人活过来，起死回生
② enjoy by young and old：被所有人喜欢

N: 在圣诞节的时候，许多教堂都举行关于耶稣诞生的圣诞剧演出。他是真的出生于一个马厩里吗？

T: 没人知道。有人认为他出生在一个牛栏里，而另一些人认为是在一个洞穴中。现在，无论是哪种情况，现在在伯利恒没有留下任何东西。

N: 他在那里生活了多久？

T: 只待了两年，他的父母就逃到埃及，以避免国王 Herod 的迫害。

N: 我记得，Herod 已经听说了一个国王诞生在这里了，所以他决定杀死他的潜在对手，尽管耶稣只是一个婴儿。

T: 后来，在国王 Herod 死后，耶稣回到了以色列，全家人住在拿撒勒。

N: 在那里他有了几个兄弟姐妹。

T: 不错，然后在他30岁的时候，开始在加利利海周围公开布道。

N: 这就是为什么他的早期弟子中有那么多渔民，像彼得和詹姆斯以及约翰。

T: 并且他做了很多不可思议的事，像同时给5000人赐食，把水变成酒，和在水面上行走等等。

N: 别忘了还有医治盲人、瘸子和生病的人。但我觉得他最大的奇迹

是让拉撒路起死回生。

T: 他是一位伟大的导师，他讲的很多故事每个人都爱听，像"浪子回头"以及"丢失的羊"的故事。

N: 我想最大的悲剧就是，绝大部分他自己的民众，犹太人，绝不相信他是上帝的儿子。

T: 犹太人期待着弥赛亚的到来，但他们盼望的是一个能使他们摆脱罗马征服者的人，而不是一个将要被钉死在十字架上人。

N: 我们基督徒相信耶稣是为我们所有的罪孽而死的，有一天他会回来结束所有的战争和苦难。

T: 犹太人相信耶稣是一位伟大的先知，但那些称他为上帝的人是对他的亵渎。

N: 我理解。我想我们最应该做的就是要了解彼此的宗教，然后求同存异，对吧？

T: 说得好！

Questions 2

1. What do you know about the life of Jesus?
2. Do you think it possible that he is God? Why? Why not?

Dialogue 3

May: Can you tell me why is it that if Christians believe in peace and goodwill on earth to all men have there been so many wars started by Christians?

Nick: Good question! Let me answer that by saying that many people start wars with more than religion on the agenda.

M: What do you mean by that?

N: Some people have power, land or wealth as their primary goals and they may use religion **as a cloak for their activities**[①].

M: I see. But it still does not explain wars like the Crusades and the Gulf wars.

N: The Crusades were about recapturing Jerusalem from Arabic hands and placing them into Christian ones. The Gulf wars were about getting rid of a tyrant and enabling the Iraqi people to choose their own government.

M: I think it was all about oil. Why can't countries like America stay out of other country's internal affairs?

N: The *Bible* teaches us that everyone is our neighbour and that we should desire people's highest good. How can a person stand by as his neighbour suffers?

M: But that's my point! Christians say that charity begins at home so why can't you Christians build a harmonious society just as we Chinese are doing?

N: Because we now live in a multi-faith world. America and the U.K. are not 100 percent Christian countries and we can't enforce our beliefs on other people.

M: But you enforce your tanks and soldiers on them!

N: I agree with you. Wars don't settle anything.

M: They say that for the last 100 years there have only been about 40 days when there wasn't a war somewhere.

N: Christians believe that one day Jesus will return and put an end to all wars.

M: Well, he should hurry up!

习惯用语 3

① as a cloak for sb.'s activities：掩盖某事

M: 你能告诉我，如果说基督徒相信世界上所有的人享有和平和善意，那么为什么这么多战争是由基督徒发起的？

N: 这是个好问题！我得说，很多人发动战争的动因超出了宗教的范畴。

293

M: 你指的是什么？

N: 有些人把拥有权力、土地或财富作为自己的首要目标，他们可能利用宗教作为掩护。

M: 我明白了。但这仍然不能解释像十字军东征和海湾战争这样的战争。

N: 十字军东征是要从阿拉伯人手中重新夺回耶路撒冷，并把他们变成基督徒。海湾战争是为了除掉一个暴君，让伊拉克人民选择自己的政府。

M: 我认为是为了石油。为什么像美国这样的国家不能对其他国家的内政置身事外？

N:《圣经》教导我们，每个人都是我们的邻居，因此我们应该期望人们最大的幸福。试问一个人怎能在他的邻居受难之时袖手旁观呢？

M: 但这是我的观点！基督徒说，慈善始于家，因此为什么你们基督徒不能建立一个和谐的社会，正如我们中国人正在做的一样？

N: 因为我们现在生活在一个多信仰的世界里。美国和英国都不是百分之百的基督教国家，因此我们不能给其他人强加我们的信仰。

M: 但你们强加给他们坦克和士兵！

N: 我同意你的观点。战争解决不了任何事情。

M: 据说在过去 100 年间，在世界各地都没有战争的时间只有大约 40 天。

N: 基督徒相信，有一天耶稣会回来，并结束所有的战争。

M: 好吧，他应该赶紧了！

Questions 3

1. How can we put a stop to wars?
2. Should countries interfere in other countries affairs? If yes, why? If no, why not?

图书在版编目（CIP）数据

嘻哈口语. 你说东，我说西/（英）斯特克（Stirk，N.）
编著. —北京：外文出版社，2007
（英语国际人）
ISBN 978 - 7 - 119 - 04874 - 1

Ⅰ. 嘻… Ⅱ. 斯… Ⅲ. 英语–口语 Ⅳ. H319. 9

中国版本图书馆 CIP 数据核字（2007）第 133442 号

英语国际人

嘻哈口语你说东，我说西

作　　者	Nick Stirk（英）	
翻　　译	张满胜	
策　　划	蔡　箐	
责任编辑	李春英	
封面设计	红十月设计室	
印刷监制	冯　浩	

Ⓒ外文出版社
出版发行　外文出版社
地　　址　中国北京西城区百万庄大街24号　　邮政编码　100037
网　　址　http://www.flp.com.cn
电　　话　（010）68995964/68995883（编辑部）
　　　　　（010）68320579/68996067（总编室）
　　　　　（010）68995844/68995852（发行部/门市邮购）
　　　　　（010）68327750/68996164（版权部）
电子信箱　info@flp.com.cn/sales@flp.com.cn
印　　制　北京飞达印刷有限责任公司
经　　销　新华书店/外文书店
开　　本　大 32 开　　　　　　　　　　印　　张　9.5
印　　数　00001 - 10000 册　·　　　　　字　　数　200 千字
装　　别　平
版　　次　2007 年第 1 版第 1 次印刷
书　　号　ISBN 978 - 7 - 119 - 04874 - 1
定　　价　18.00 元